PM 645
Centre for Mathematics Education
The Open University in association with
The Inner London Education Authority

Girls into mathematics

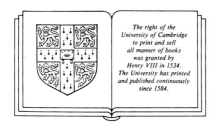

Cambridge University Press

Cambridge

London   New York   New Rochelle

Melbourne   Sydney

Writing team
Leone Burton, Formerly Avery Hill College, ILEA; now Thames Polytechnic
Pat Drake, ILEA teacher
Judy Ekins, Mathematics Faculty, The Open University
Lynne Graham, Mathematics Faculty, The Open University (Chair/academic editor)
Margaret Taplin, Mathematics Faculty, The Open University
Gaby Weiner, School of Education, The Open University

Acknowledgements
During the production of this pack we worked with groups of teachers in London,
Hertfordshire and Wakefield. The response to our request for help was overwhelming.
Nearly 100 teachers participated in the development of the materials, and we would like to
express our sincere thanks to those who were involved in trying out and generating ideas
as their comments were vital in shaping the final product. Indeed the pack would not be as
it is without their help. We would also like to thank colleagues from the Open University
and from the ILEA for their comments on draft versions of the text, and also the secretaries
and typists in the Mathematics Faculty at the Open University, who typed the materials.

The Open University
Walton Hall
Milton Keynes MK7 6AA
in association with the Inner London Education Authority

Further information on this and other Open University courses may be obtained from the
Learning Materials Services Office, Centre for Continuing Education, The Open
University, PO Box 188, Milton Keynes MK7 6HN.

Published by the Press Syndicate of the University of Cambridge
The Pitt Building, Trumpington Street, Cambridge CB2 1RP
32 East 57th Street, New York, NY 10022, USA
10 Stamford Road, Oakleigh, Melbourne 3166, Australia

First published 1986

Printed in Great Britain by Scotprint Ltd., Musselburgh

British Library cataloguing in publication data

Girls into mathematics.
1. Mathematics – Study and teaching – Great Britain
2. Sex discrimination in education – Great Britain
I. Open University   II. Inner London Education Authority
510'.7'1041   QA14.G7

ISBN 0 521 31094 6

Contents

Foreword

Many teachers are currently concerned about equality of opportunity in schools. The ILEA has equal opportunities policies concerning sex, race and class, and is currently asking all educational institutions to discuss these policies and to draw up programmes for their implementation. These policies are likely to be 'whole institution policies' or 'whole school policies' to which all teachers and pupils subscribe. One aspect of equal opportunities concerns the curriculum, including subject content, presentation and distribution of resources as well as matters such as teaching method, classroom organisation and pupil evaluation. In secondary schools, subject teachers need to examine closely how issues of equal opportunities relate to their own specialism. An issue of major concern is the general under-achievement of girls relative to boys in mathematics. This pack seeks to inform teachers about the issue and to suggest strategies for classroom action. The pack presents research findings, suggests activities based on the findings and then invites comment on the activities' results.

Unfortunately, there has been very little research in relation to girls and mathematics which takes racial, cultural or class differences into account. This poses a problem. Either the pack could be held up until the necessary research is done – which could take some time – or it could be issued as it is without detailed evidence on the influence of race and class. The pack has been published since we feel that its use encourages a better understanding of the needs of *all* children. The main thread of the pack is the gender issue, since all the activities have been developed from that point of view. However, interwoven with this are frequent comments inviting teachers to re-examine their results from ethnic, cultural and economic viewpoints. In this way, we hope that individual teachers and whole departments will be encouraged to carry out their own small-scale research on the effects of race and class as well as gender, and this might point the way to the future research work which needs to be done.

This package has come about through collaboration between the ILEA and the Open University, and breaks new ground in Britain. It is the first in-service material designed to encourage teachers to reflect seriously about their attitudes to girls and boys and mathematics.

Carol Adams   Inspector for Equal Opportunities, ILEA
Hugh Neill   Staff Inspector for Mathematics, ILEA
November 1985

Study guide

Ways of using this pack

This Open University/Inner London Education Authority pack is designed to be studied by teachers. It is likely that the materials will be used by Local Education Authorities as part of their in-service provision, but they can also be studied as a school-based in-service course, with one member of staff taking responsibility for the organisation. (If you are interested in running such a course, we have produced a booklet entitled *Notes for running a study group*, which contains some suggestions on how to organise a series of workshops based on the pack. This booklet is available from the Learning Materials Services Office at the Open University.) Alternatively, the pack can be studied by individual teachers since, in common with other Open University materials, it is designed for self-study.

(*Note*. Although this pack originated (partly) from the Open University, it does not form part of the OU undergraduate programme and its use by teachers is not necessarily organised or accredited through the Open University system. This explains why it is termed a 'pack' rather than a 'course'.)

Organising your time

The pack consists of five chapters and an appendix, each of which is divided into subsections. We start each subsection by establishing the existence of specific difficulties faced by girls. We do this in two ways:

  (i) by summarising the available research evidence; and
  (ii) by inviting you to examine your own school structure and teaching practice.

We then consider how the various problems can be wholly, or at least partially, resolved.

It should take you about 40 hours to work through the pack. We expect you to be actively involved throughout your study and to this end the text contains the following different kinds of activities; the symbols by each activity show you what to expect.

Here you are asked to jot down ideas, to reflect on something you have read or to analyse some data. These activities should involve only a few minutes' work.

Here you are asked to try out an idea, to make observations or to collect your own data. Frequently these activities will involve you in working with pupils in the classroom, though occasionally you will be asked to examine classroom materials or other information obtainable from school. Most of these activities require prior organisation and will take some time to carry out.

Here you are asked to compare your experience and/or views with those of your colleagues in order to examine a problem in a context wider than your own classroom. Many of these activities also require prior organisation.

Each activity in the text is followed by comments based on the feedback from teachers involved in developing this pack or on relevant research work.

We suggest that you begin your study of each chapter by reading it through and by trying all the 'pause for thought' activities. As you are working through the material, make a note of the 'school-based activities' and 'discussion topics' that are most relevant to your school or your teaching, and try to plan when you can fit them into your workload at school. It is not essential to carry out these activities at the exact point where they appear in the text, but you should aim to do them round about the time that you are studying the relevant material.

(*Note*. You might find it useful to contact one or two sympathetic colleagues *before* commencing your studies, so that you can meet them regularly for the 'discussion topics'. Teachers involved in testing the activities reported that it was much easier to talk with people who were already thinking about some of the issues.)

Adapting activities to suit your circumstances

We believe that this pack is relevant to all teachers of mathematics. However, you may well find that some of the activities do not immediately appear to be suited to your school and to your experience. For example, several activities ask for comparisons between girls and boys, and these may not seem relevant in a single-sex school. However, many teachers from single-sex schools observed that the range of roles taken by girls and boys in mixed-sex classes were also adopted within the one sex in their classes, so any comparisons could be carried out from this point of view. The issues themselves are, of course, always pertinent and our comments suggest ways of reflecting on your own results whatever your teaching situation. You should be able to use many of the activities as written; alternatively, you might want to adapt some of them slightly so that you are comparing, for example, different age-groups rather than different sexes.

Similarly, although many of the activities were originally designed for secondary-school teachers, you should find that they are equally relevant to teachers of other age-groups. Again you may be able to use the activities as written, or you might prefer to adapt them slightly to suit your individual circumstances. For example, whenever we refer to specific mathematical content, you could adapt it so that it is more suited to the abilities of your pupils, yet still retaining the same characteristics.

Of course, your results may not necessarily agree exactly with our comments. For example, girls in single-sex schools may do better and have more positive attitudes than girls in mixed-sex schools, and it may be harder to discern differences in mathematical performance between girls and boys in their early years than it is in their adolescence. Nevertheless, we hope that this pack contains useful advice and stimulation for thought whatever your teaching situation.

Issues of race and culture

Although we have concentrated in this pack on the problems of girls who, *in general*, under-achieve relative to boys in mathematics, it should not be thought that issues of gender are the only ones which affect achievement or under-achievement in the subject. Race and class also have an effect. For instance, there is some evidence that Afro–Caribbean girls perform better than other girls in mathematics, and that Asian boys perform worse than other boys (see *Educational Studies in Mathematics* **14**, November 1983, p. 342). This suggests that the influence of racial and cultural background could be no less pervasive than that of gender.

Unfortunately, although gender-based research into learning mathematics exists – even if it only gives indications of possible reasons for success and failure – research into the varieties of experience related to race, culture and class background is virtually non-existent. And those few studies which have been carried out have concentrated on general educational performance rather than mathematics in particular. Consequently the research quoted in this pack does not always take into account the dimensions of race and cultural background, although it is argued that this lack of 'hard' evidence needs to be redressed and it is hoped that future research into achievement in mathematics will examine both gender and cultural background.

All the activities in this pack have been carried out by teachers – both black and white – in mixed-race classrooms, but since the activities are concerned with gender issues, the feedback that we have received has been expressed mainly in these terms. However, in using the pack, you may find it useful and relevant to take account of the racial and cultural dimensions of the gender issues, and appropriate suggestions are made throughout the pack.

Following up the research: accompanying book

This pack summarises much of the existing research evidence on girls and mathematics, particularly that which is relevant to Britain, but you may well wish to follow this up in greater detail. At the end of each chapter you will find a list for further reading. The bibliography at the end of the book supplies details of the books and articles mentioned in the text.

The most commonly-referenced articles have been collected together into a single book, *Girls into maths can go*, edited by Leone Burton and published by Holt, Rinehart and Winston.

Introduction

0.1 What are the problems?

My youngest child has just started school and I'm now looking for a job. I've been for one interview, but it really put me off because I found out that I'd be expected to work with figures . . . Maths always terrified me at school even though I managed to get a good grade at CSE . . . (Lorraine)

I was OK at maths, but I only studied it up to O-level then I went on to do other things – English and history . . . In our school it was mainly the boys who took A-level maths; I never really thought about it. Looking back though, I wish I'd done more because I use a lot of statistics in my job and I don't have a clue about what I'm doing half the time . . . I always seem to feel out of my depth . . . (Yvonne)

I've got an honours degree in mathematics and I'm a teacher in a large comprehensive. I think I'm a fairly good teacher – at least up to 16. I've never taught at A-level although I'd like to . . . Perhaps they don't think I'm good enough – maybe that's why I seem to get left out of staffroom conversations about how to teach calculus and so on . . . or perhaps it's just because all the rest of the department are men! (Sharma)

I work in a department where there are several women – all with responsible jobs – all mathematicians. They're just as well qualified as the men, sometimes better qualified, yet there are no female project leaders. This is not due to lack of opportunity . . . I think it's because the women don't want the responsibility. They never seem to like taking decisions on their own . . . perhaps they just lack confidence. (Dave)

I wasn't very good with numbers when I was in school; in fact I only managed to scrape an O-level at the third attempt – after a lot of coaching! . . . I needed the qualification in order to become an engineer . . . Now I'm fortunate enough to be in a position of getting other people to do all my calculations. But I've always got a good idea of what's going on and the sorts of results I'm expecting, and that's the important thing . . . that's why I've been promoted so many times. (Sean)

These anecdotes exemplify some of the different views that men and women hold about themselves with regard to mathematics: women are more likely than men to consider themselves as mathematical failures, whether or not this is justified; they are more likely to regret their lack of success and to feel hampered by it in their careers; they are more likely to feel that their mathematical interests were discouraged in one way or another at school. Even when women are extremely successful in terms of qualifications, they tend to hold fewer posts of responsibility, and to be less confident than their male counterparts.

You might say that these are typical adult viewpoints, formed on the basis of past experience. Perhaps it is the case that in the past boys were more successful than girls, that they did like the subject better, that they did receive more encouragement at school etc., but surely changing attitudes mean that today's schoolgirls have more positive self-images? Well, the question of girls' under-achievement in mathematics provided the focus for an appendix in the Cockcroft report *Mathematics Counts* (1982) and it is now a well-recognised issue of public concern. Yet evidence suggests that in general girls *continue* to be less successful than boys in mathematics and that they still tend to have negative feelings about the subject. For example, figure 0.1 provides the most up-to-date information available at the time of publication about girls' achievement relative to boys in public examinations in England.

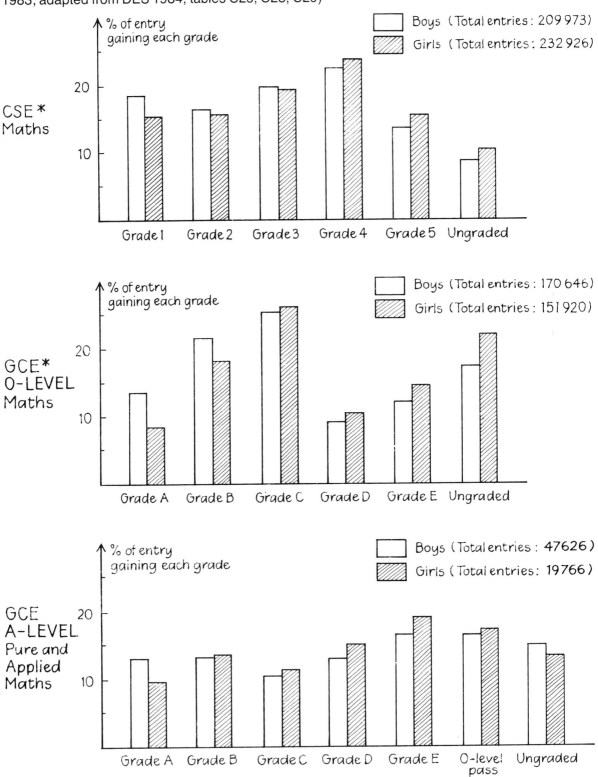

Figure 0.1 Public examination results in mathematics (England only; summer 1983; adapted from DES 1984, tables C25, C28, C29)

* includes entries for joint CSE/GCE 16+ feasibility and development studies

Activity 0.1

How do the statistics in figure 0.1 compare with those for your school (or Local Authority if the data are available)?

(i)   Record the total numbers of boys and of girls entered for each type of public examination last summer (CSE, O-level, 16+, A-level etc.) in your school or area.

(ii) For each public examination, calculate the percentages of each sex gaining each of the possible grades. Prepare comparative bar charts like those in figure 0.1.

(iii) What evidence is there (if any) that girls are under-achieving in mathematics? Compare your results with the DES (Department of Education and Science) statistics in figure 0.1.

Comment

The DES figures indicate the following:

■ About the same numbers of girls and boys were entered for public examinations in mathematics at 16+. However, a larger proportion of boys were entered for GCE O-level (45% of the total boys' entries for GCE/CSE were for GCE O-level, compared with 39% of the total girls' entries).

■ More than twice as many boys as girls were entered for GCE A-level.

■ The boys performed better than the girls in all three public examinations. The differences were small at CSE, but were more marked at GCE O-level and A-level. Notice in particular the relative percentages of girls and boys gaining the top grades in all these exams.

Your school or Local Authority statistics will probably reflect this trend even though the DES figures relate to 1983 and apply only to England. Of course, there are schools – and Local Authorities – where the girls perform better than the boys. If your results suggest that this is happening in your area you may like to consider why.

You may also like to investigate whether there are any differences in the performance of pupils from different cultural or class backgrounds.

Examination statistics confirm that on the whole boys continue to perform better than girls in mathematics at 16+, yet there is no evidence that boys are *more able* than girls; indeed girls perform better than boys in many other subjects at this level. Furthermore, the fact that there are schools where girls and boys are equally successful in maths examinations (for example, in one of the Wakefield schools which was involved in the development of these materials) suggests that girls *in general* could achieve more than they are doing at the moment. In other words, it seems that many girls are currently *under-achieving* in mathematics. Why is this?

And why do so many girls (even when they have ability in the subject) choose not to study mathematics when they have the choice? Such 'opting out' is readily observed in the sixth-form, but it also occurs lower down the school. Maths is a compulsory school subject up to the age of 16 so British girls cannot opt out directly (as happens in the USA); however, they can and do opt out *indirectly* since low attainment can be a form of learning refusal. Girls are also less likely than boys to choose traditionally maths-related subjects in their courses of study up to 16 (see figure 0.2). Since mathematics is becoming increasingly important to everyone in everyday life and work, the number of students opting out of its study – at any level and in any form – is disturbing.

Figure 0.2 Entries at O-level/CSE in selected subjects (England only; summer 1983; adapted from DES 1984)
The numbers of entries in English, a compulsory school subject, provide a basis for comparison.

Subject	Total boys' entries	Total girls' entries
English	569 695	607 952
Physics	257 868	85 129
Chemistry	152 740	110 945
Computer studies	54 412	28 766
Total science and technology	1 519 049	1 041 036

The fact that the proportion of girls doing so is greater than that of boys explains why this pack is entitled 'Girls *into* mathematics'!

The pack thus addresses two principal questions:

- Why do girls tend to under-achieve (relative to boys) in mathematics?
- Why do so many girls choose – directly or indirectly – not to study mathematics or maths-related subjects?

The answers to these questions are obtained by investigating the many factors which can inhibit the mathematical performance and affect the attitude of any student, male or female. It turns out that girls are more susceptible than boys to many of these influences.

Activity 0.2

What are the factors that could particularly inhibit girls' performance in mathematics?

(i) Find out the views of your colleagues by asking them to complete the sentence 'In my school a reason why girls under-achieve or opt out of mathematics is . . .'

(ii) Compare the responses.

Comment

Try to group your findings along the lines of figure 0.3. If necessary, add your own points to the diagram (for example, you may wish to include specific points regarding cultural and class background).

Of course, there is a considerable overlap between the factors identified in figure 0.3, and it is more likely to be a combination of several or all of them (rather than any particular one) which contributes to the problems experienced by girls in mathematics. However, a closer look at each issue will lead, it is hoped, to a greater understanding of the total problem.

0.2   What to expect from this pack

This pack is divided into five chapters and an appendix:

Chapter 1   Mathematics and the curriculum
Chapter 2   Feelings, attitudes and expectations
Chapter 3   In the classroom
Chapter 4   Bias in teaching materials

The appendix is included because of the many real or apparent connections between mathematics and computing. For example, many teachers have responsibilities for both mathematics and computing, school computers are frequently sited in the maths department, and computers are often used more in maths classes than in other lessons. These connections may partly explain why girls are already developing negative attitudes towards computers even though the technology has only relatively recently been introduced into schools. For these reasons it may be desirable to sever the links between the two areas. On the other hand, microtechnology is having a very influential effect on the teaching of mathematics and we argue that *strengthening* the links could have a positive effect on girls' attitudes towards both the computer *and* mathematics.

Within each chapter we discuss specific problems faced by girls when studying mathematics and consider how these can be wholly – or at least partially – resolved. Often we suggest action which can be taken in the classroom or school, although there are some instances where there is no easy solution other than being aware and reacting sensitively. Do not be surprised or disappointed at your colleagues' reactions to some of the ideas put forward: anti-sexism has gone a long way beyond the niceties of who opens the door, but some of the issues may still be ridiculed by some people. Our aim is to provide you with facts, to help you contend with adverse reactions if and when they occur.

Figure 0.3   Factors which might particularly inhibit girls' mathematical performance.

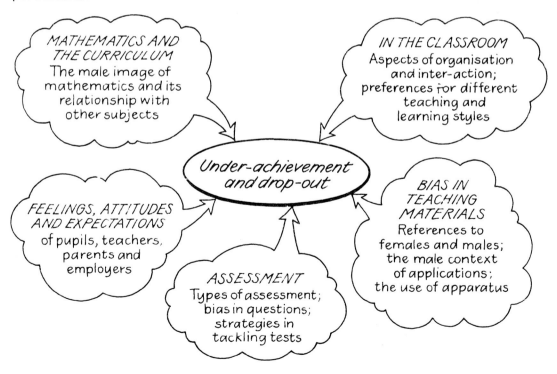

On a wider front, we hope that by concentrating on the problems of *girls*, you will find that you increase your understanding of *all* the pupils that you teach, since they are all individuals with differing needs. Improving the situation for 52% of the school population should affect the learning experiences of all pupils.

You can also use the pack to build up a better rapport with girls and boys from all backgrounds. During the development of these materials we discovered that pupils of both sexes were extremely interested in discussing the issues raised. Although some of the 'school-based activities' provide opportunities for pupils to air their views, you may wish to involve them on a more regular basis; for example, the 'discussion topics' are often just as relevant to young people as they are to teachers, and you may find that your pupils want you to report your results and research findings to them. Many of the teachers who have already worked through the pack said that they gained greater knowledge and understanding of their pupils and that the classroom atmosphere improved noticeably as a result of such discussions.

When carrying out observations which form the basis for your research, bear in mind that it is not enough to notice something just once; for example, that you tend to respond to pupils' questions differently, and that their gender and/or race affects the response. Such an observation needs to be intentionally looked out for, lesson after lesson. Very brief anecdotes ('I had just answered a question from Ravi and was about to respond to Yvonne, when I suddenly noticed . . .') can be collected and shared with colleagues. Such sharing is analogous mathematically to collecting particular cases in which a general pattern is sought. The purpose of sharing is to seek resonance with other people's experience and to sharpen your own awareness, so that when the need arises you have access to a broader range of options than simply your automatic reaction.

We very much hope that you enjoy studying this pack, although remember that 'what you get out of it depends on what you put into it'. It is strongly recommended that you work through as many of the activities as possible. Not only should they help you to test the ideas contained in the text, but they should also help you to turn ideas into experience and experience into practice.

1 Mathematics and the curriculum

> Decimals for instance and fractions, I don't know I just didn't seem to grasp them. I'm the kind of person that needs to have things spelt out very clearly and I just didn't . . . I found them boring. I couldn't concentrate on them at all. They weren't interesting enough for me. It wasn't attractive enough.
>
> (quoted in Graham and Roberts 1982)

This is how one woman who took up the study of mathematics in adulthood described her memories of school mathematics. For her, school mathematics had a poor image and, even in this short statement, she implies a number of different reasons for this:

- the mathematics topics were too difficult;
- the way she was taught did not allow for sufficient time or discussion;
- mathematics, as a subject, lacked interest – it was boring;
- mathematics seemed irrelevant to her life.

There might have been all sorts of reasons why the woman felt this way, some of which probably related to her own schooling experience – for example, she might have been subjected to old-fashioned teaching and/or textbooks, a rapid turnover of staff, lack of school resources and equipment, and little support from home. Yet many adults, both women and men, see mathematics in much the same sort of way (see Buxton 1981), and there is considerable evidence that many pupils hold similar views.

Activity 1.1

What are the views of your pupils? Give a copy of figure 1.1 to each person in one of your classes. Have some extra copies for when the class divides into groups at stage (ii).

Figure 1.1 Ideas about mathematics

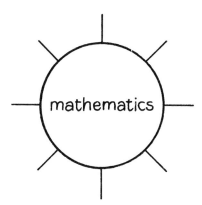

(i) Ask pupils to write against each spoke on the diagram a summary of the ideas they have about mathematics. (For example, they might find geometry – or even all mathematics – 'interesting'.) Emphasise that for the moment you are interested in their *own* views. (You might like to compile a diagram of your own at the same time.)

Figure 1.2   One girl's image of mathematics

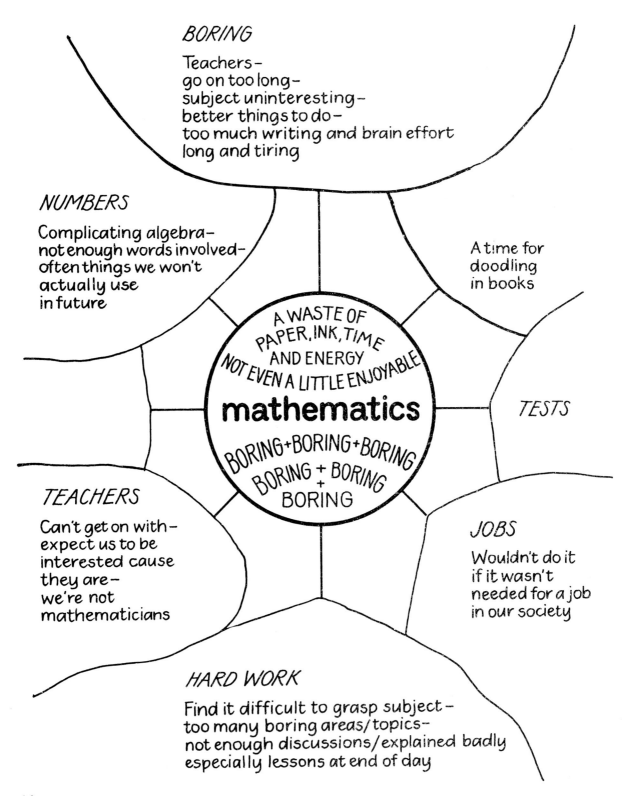

(ii) When this exercise has been completed, direct pupils to form groups of three or four to compare what each of them has written. Ask them to produce on a clean diagram an agreed summary of all their individual ideas.

(iii) Collect the group diagrams and count the number of times that various ideas are mentioned. Discuss the results with your pupils. Can you identify particular aspects of mathematics which seem important to your pupils? Are you at all surprised by your results?

Comment

A class of fourth-year girls from an all-girls' secondary school in London carried out this activity. According to their teacher, they were generally of average ability and normally enthusiastic and hard working, yet figure 1.2 demonstrates that at least one of these girls held very negative views about mathematics. Figure 1.3, which summarises the ideas mentioned in the ten group diagrams, suggests that such views were shared by many of her peers.

Figure 1.3 Summary of pupils' ideas about mathematics

Topic	No. of times mentioned	Topic	No. of times mentioned
Difficult/hard work	12	Calculations	4
Boredom	8	Four rules	4
Teacher	6	Frustration	4
Examinations	6	Interest	2
Numbers/figures	5	Equations	2
Fractions	5	Vocational need	2
Algebra	5	Calculator	2

Topics mentioned *once* were: decimals, trigonometry, percentages, π, angles, area, homework, formulas, doodling, maths as a boys' subject.

Were your results similar? Did you find that your pupils also have a narrow and school-defined perception of mathematics? Were there any differences in the views expressed by girls and by boys, or by pupils from different cultural and class backgrounds?

This chapter considers why mathematics, as a subject, has such a poor image with so many people – especially with girls. We begin by exploring the place of mathematics in the school curriculum, to see whether particular subject combinations or cross-curricular practices have implications for girls' views and mathematical performance. We then look at the historical development of mathematics as a school subject and as an academic discipline to which few women could gain access. We discuss the struggles of women mathematicians in the past and consider whether some of the barriers they faced still exist for girls (and some boys) today. The final section looks at the present mathematics curriculum; in recent years there have been a number of changes – could any of these help to improve the situation for girls?

1.1 Mathematics: its relationship with other subjects

This section follows up the brief comment in the Introduction about the way in which many girls opt out of mathematics indirectly by choosing not to study maths-related subjects.

Activity 1.2

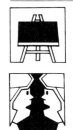

Look at the table in figure 1.4.

(i) In column (1) mark with an asterisk (✳) the subjects that you consider to be maths-related.

(ii) In column (2) try to categorise each subject as a 'girls' subject' (G) or a 'boys' subject' (B). Bear in mind the figures given on entries and exam performance as well as your own views. If there appears to be no particular bias, leave a blank.

(iii) Record your results in the table in figure 1.5.

(iv) Compare your results with those of your colleagues; for example, people have different views about what constitutes a maths-related subject.

Figure 1.4 Entries at O-level/CSE in selected subjects (England only; summer 1983; adapted from DES 1984)

(1) ✳	(2) G/B	(3) Subject	(4) Total number of entries		(5) % of entry awarded O-level grades A–C/CSE grade 1	
			Boys	Girls	Boys	Girls
		English	569 695	607 952	27.82	36.37
		Mathematics	380 619	384 846	37.57	30.25
		Religious studies	52 033	79 926	31.87	39.74
		History	146 913	155 840	32.83	32.77
		French	119 600	195 766	38.53	39.68
		German	30 823	52 286	42.25	45.77
		Spanish	5 890	12 647	48.00	49.93
		Music	10 074	19 656	41.65	48.17
		Art/Craft	150 421	163 867	33.99	42.63
		Computer studies	54 412	28 766	39.29	28.49
		Physics	257 868	85 129	39.11	42.91
		Chemistry	152 740	110 945	43.83	39.85
		Biology	145 455	296 352	38.97	30.74
		General science	49 978	37 762	8.44	8.74
		Geology	11 810	4 761	43.62	44.89
		Technical drawing	138 696	8 303	29.00	24.99
		Building/Engineering studies	39 784	1 145	20.74	10.57
		Agric/Hort/Rural	12 327	6 706	14.43	24.01
		Woodwork	68 624	1 769	20.05	16.62
		Metalwork	83 868	732	19.94	15.98
		Economics	33 086	24 697	46.58	42.42
		Social studies/Sociology	57 069	93 141	19.18	30.14
		Geography	231 291	171 864	34.60	34.55
		Needlework	140	44 252	30.00	30.17
		Other domestic subjects	16 780	187 937	8.59	27.00
		Accounts	9 593	15 457	47.82	47.76
		Commercial subjects	29 461	75 151	32.08	24.64

(*Note.* The table includes pupils who sat joint 16+ criteria exams.)

Figure 1.5

Category	Number of subjects
✳ G	
✳ B	
G	
B	
Blank	

Comment

Teachers in the workshops indicated that in general:

- there are more maths-related 'boys' subjects' than maths-related 'girls' subjects';
- there are very few 'boys' subjects' with no mathematical content;
- a comparatively high number of 'girls' subjects' contain little or no mathematics at all.

Did you reach similar conclusions? Why do you think this is the case? Were you surprised by any of the statistics in figure 1.4 (for example, by the exam performance in physics)?

Activity 1.2 confirms that in England girls are less likely than boys to choose to study maths-related subjects – perhaps because those subjects which have traditionally been maths-related tend to be classified as 'boys' subjects'. For example, figure 1.4 suggests that physics, chemistry, geography and the technological and craft subjects are taken by more boys than girls whereas it is the other way round in languages, biology and domestic subjects. Admittedly, there may be some variation in the pattern of subject take-up – for instance, in your school do you notice any differences in the preferences of pupils from different cultural and class backgrounds? Here we concentrate only on the *general* trends displayed by girls and boys as demonstrated in activity 1.2.

John Ling (1977) has explored the amounts of mathematics in school subjects across the curriculum. He claims that:

- *Physics* is by far the most extensive user of mathematics, drawing on all aspects of mathematical enquiry.
- *Chemistry* places relatively low demands on mathematical skills up to O-level; nevertheless it requires use of symbols, decimals, ratio, proportion, algebra, geometry, logarithms and so on.
- *Biology* involves a fairly small range of mathematical topics: graphical representation, percentages, proportion, scale factors, volume and statistics.
- *Geography* uses a considerable number of mathematical skills, particularly diagrammatic techniques, mappings and statistics.
- *Economics* and *social studies* involve mostly arithmetic (percentages) and graphs.
- *Technical subjects* are practical rather than theoretical but nevertheless involve arithmetic, mensuration, geometric construction and orthographic projection.
- *Domestic subjects* mainly demand competence in measuring and arithmetic.

Activity 1.3

John Ling's work was carried out in 1977. Do you think that this is a fair summary of the mathematics used across the curriculum in schools today?

Comment

Some teachers at workshops criticised Ling for failing to recognise important mathematical content in 'girls' subjects' – for example, the use of statistics in biology – and claim that this analysis is but a reflection of Ling's assumptions about the status of particular mathematical topics. Others pointed out that there is a high proportion of mathematics in (say) dressmaking and needlework, which girls and their teachers do not acknowledge. It was also pointed out that, in many ways, Ling's work still provides a fair summary of the way that mathematics is viewed in schools today; the mathematics in 'female pursuits' was (and still is) under-valued relative to the mathematics in traditional 'male subjects'. Little wonder that pupils often report that the former is 'not really maths – real maths is hard'.

The discussion so far demonstrates that boys tend to receive considerably more reinforcement of *formal* school mathematics in their other subjects than girls. There is some evidence that this additional mathematics practice contributes to boys' higher attainment in maths examinations at 16+. For example, a study by Shiam Sharma and Roland Meighan (1980) suggests that it is maths experience in other subjects, rather than the gender of the pupil, which most influences pupil performance. In their investigation of mathematics performance at O-level, they found that though, overall, girls did less well than boys, girls with a science or technical background performed as well as boys with a similar background. Conversely, no pupils in the sample without a science or technical background (whether girls or boys) obtained the two highest O-level mathematics grades. (Interestingly though, girls actually did better than boys in this category.)

There is some evidence therefore that the relationship between mathematics and other parts of the curriculum is of importance. Mathematical activities and practice, *seen* as formal mathematics in lessons of other subjects, do seem to affect pupil performance. Could it be that girls are handicapped in their mathematics because they choose not to study technical or science subjects alongside mathematics? Or could it be that mathematics is so off-putting that they are discouraged from studying other subjects which they perceive as heavily mathematical?

To summarise, it seems that if girls are encouraged to go into scientific or technological fields of study, their mathematical skills will improve. An improvement might also be effected if the mathematics inherent in many of the subjects traditionally studied by girls is recognised as part of the *formal* curriculum.

What can be done?

There are a number of ways in which such cross-curricular issues concerning girls and mathematics can be discussed, if not resolved. You might like to consider the following:

(1) You could encourage the establishment in your school of a cross-curricular mathematics working party with representatives from each subject department to discuss the mathematics used throughout the school, to negotiate an agreed approach to teaching certain mathematics topics across curricular boundaries, and to acknowledge and highlight the importance of mathematics in 'girls' subjects'.

(2) You could consider extending the core curriculum so that all pupils are compelled to take at least one other science or maths-related subject in addition to mathematics.

(3) You could attempt to ensure that the timetable is not blocked in such a way as to encourage stereotyped choices. For instance, when physics is blocked against typing, or home economics against CDT (craft, design and technology), the implications for pupils are that 'typists' and 'physicists' are mutually exclusive categories and that pupils interested in 'cooking' cannot possibly want to take up 'carpentry'.

1.2 The historical development of mathematics as a subject in the British school curriculum

Most people have been brought up to believe that 'maths is maths is maths', i.e. that mathematics, the science of space, number and quantity, is a form of *neutral* enquiry with a distinctive internal logic and an established structure. But, as we shall show in this section, school mathematics is socially produced and has changed over the years to meet perceived social and economic needs. Examining the historical development of the subject might shed some light on the problems faced by girls.

Surprisingly perhaps, mathematics was not always an important subject on the school curriculum. In the early part of the nineteenth century many of the long-established grammar and public schools were restricted by their charters to the teaching of the Classics. In the expansion and restructuring of state and private schooling in the nineteenth century, in order to deal with the enormous social and economic changes brought about by the Industrial Revolution, it was left to the newer, private schools, created by the increasingly powerful and aspiring middle classes, to establish mathematics as a school subject. However, most of these schools, whether public or private, new or old, were boys' schools. In the relatively few girls' schools of the time, arithmetic rather than mathematics was taught, though not without protests from some parents about its unsuitability as part of the education of young ladies.

The separation of mathematics and arithmetic in the school curriculum was not just a question of gender. In the clearly defined and constantly reinforced class structure of the time, the contents of the mathematics curriculum and the justification for its teaching varied according to the class of pupil for which the system catered. Even after reorganisation, many of the public schools (for the sons of the upper classes) still awarded mathematics very low priority. It was regarded as too utilitarian to be of any real value to the education of gentlemen.

The newer private schools (for middle-class children) were quicker to respond to the inclusion of mathematics as a vocational subject – for entry into the services or to professional careers. Cheltenham College's 'Modern Department', for instance, placed mathematics in a dominant position in the curriculum as a means of training boys for the military academies of Woolwich and Sandhurst, for engineering and for commercial life (Griffiths and Howson 1974). For middle-class girls, mathematics (and even arithmetic) remained an unacceptable part of their education for a longer time. In 1860 Miss Beale, Headmistress of Cheltenham Ladies College, was informed by a father, on deciding to send his daughters elsewhere,

> My dear lady, if my daughters were going to be bankers, it would be very well to teach arithmetic as you do, but really there is no need. (Kamm 1958)

However, arithmetic eventually became an acceptable subject for girls when a number of girls' grammar schools were founded in the 1870s and 1880s.

In the state elementary schools (for working-class children) both girls and boys ostensibly learned equal amounts of arithmetic. However, the time allocated was in fact somewhat different:

> . . . as the time of the girls is largely taken up with needlework, the time they can give to arithmetic is less than that which can be given by boys. (Cross 1880)

Moreover, one did not speak of 'mathematics' but 'arithmetic' as an elementary school subject. Mathematics became a distinguishing mark of male, secondary, middle-class education.

By the twentieth century the discussion had moved away from whether and to whom mathematics should be taught towards considerations of content and quantity. It was claimed that girls needed to be taught a different kind of mathematics from boys, according to their perceived biological, psychological and social differences. So for arithmetic it was stated:

> a course suitable for boys often requires considerable modification if it is to serve the needs and interests of girls . . . [Girls should deal with] detailed accounts accompanying shopping and housekeeping . . . [while boys should] establish by experimental methods some of the more important theories of elementary geometry. (quoted in Cockcroft 1982)

Activity 1.4

Can you think of similar explanations used today for why girls cannot or will not do well at mathematics?

Comment

Today it is sometimes still claimed, for instance, that:

- girls have less mathematical ability than boys because the distribution of ability in spatial skills favours males;
- girls and women are handicapped in their mathematical thinking because they are less rational and/or less logical than their male counterparts;
- since most girls become wives and mothers, there is no point in them studying mathematics beyond a certain level.

You can probably think of other claims as well. Yet there is no evidence that either of the first two points mentioned above are true. The third is a matter of opinion since it fails to recognise the fact that most mothers work outside the home and also ignores the crucial role that parents play in the early development of mathematical skills in their children (see Graham 1985).

The importance attached to domestic subjects to fit girls for their future roles as wives and mothers had implications for maths (and science) teaching throughout the first half of the twentieth century. For example, in 1923 it was recommended that girls could opt out of mathematics and science, in favour of domestic subjects.

> An approved course in a combination of these [domestic] subjects may for girls over 15 years be substituted partially or wholly for science and for mathematics other than arithmetic.
>
> (quoted in Cockcroft 1982)

Hence the principle of mathematics for boys and arithmetic for girls extended into the twentieth century, as part of the 'natural' order of things, though it was firmly emphasised that while the education of girls and boys might be necessarily different, it was to be judged equal in status and level.

The 1944 Education Act enshrined the 'different but equal' principle in the establishment of the tripartite system of grammar, technical (or central) and secondary modern schools. At this time there was no official differentiation in the schooling of girls and boys – the types of schooling pupils received depended on their perceived cognitive type, i.e. academic, technical or practical. Yet girls in some areas had to attain higher marks to qualify for grammar school places (Weiner 1985). If the same cut-off mark had been used for both sexes many more girls than boys would have been educated in grammar schools and this was not acceptable!

Stereotyped assumptions continued to limit the schooling of many girls. This was particularly evident in the type of mathematics that was taught.

Activity 1.5

Figure 1.6 (page 24) contains parts of the contents lists of three mathematics textbooks published about the same time. These provide an indication of the mathematical content seen as most appropriate for different kinds of pupils.

(i) When do you think the books were published?
(ii) Try to identify the audiences for whom the books were written.
(iii) What indication do these contents pages provide of the perceived role of school mathematics at that time?

Comment

As indicated, the audiences for whom these books were written were somewhat different. Book A is entitled *Essential everyday arithmetic for girls, part 2*, and it was written specifically for use in girls' secondary modern schools. It had a strong practical orientation and, according to the authors, Bertram and Evelyn White,

> it contains all that is necessary to enable a girl to develop facility in dealing with the various types of arithmetic which enter into her everyday life. Consequently when she leaves school she will be fully competent to deal intelligently and confidently with the wide range of practical problems based on calculations and the expending of money which are particularly associated with the home and general living.
>
> (White and White 1951)

In contrast, Book B, *Daily life mathematics, book 2*, was written for *all* pupils in secondary modern schools. However, the contents page suggests that it was aimed primarily at boys, and this is confirmed perhaps by the following extract from the preface to the course:

> The course is built round topics and projects, each of which contains a core of mathematical ideas arranged as progressive studies in the different years. These topics may be grouped under two main headings, those concerned with financial transactions and those concerned with space and time. Among the former are such subjects as Wages and Salaries, House Purchase, Endowment Insurance, Local and National Finance, Imports and Exports, Balance of Trade, the Settlement of

23

Figure 1.6   Contents lists of three school textbooks

Book A

CONTENTS

Book B

CONTENTS

Book C

CONTENTS

International Accounts. The space and time topics include Workshop Drawing, Practical Surveying, Earth Measurements, Simple Astronomy, the Construction of Mercators' Charts and their use in Air Navigation. (Burns 1952)

Book C, entitled *School Mathematics: a unified course, part 3* (Parr 1951), was aimed at grammar-school pupils (girls and boys) who were being prepared for the new O-level examination.

The date of publication for books A and C was 1951, book B was published in 1952. Incidentally, there were no books written specifically for boys during this period, but there were several series aimed at girls. It is tempting to conclude that the maths books for girls and boys were considered inappropriate for the particular (domestic) needs of some girls!

These contents pages tend to confirm that in the mid twentieth century it was considered more important for girls to study arithmetic whereas boys were encouraged to study a wider range of mathematical topics. Book A addresses topics (mainly arithmetic) *only* in domestic terms; book B, although emphasising practical applications (almost all male), focuses on some more abstract mathematical topics in addition to arithmetic; book C appears to be almost entirely abstract (and where there are references to practical applications these are again in male-dominated occupational areas).

The books also indicate that the arithmetic/mathematics class dichotomy was still evident in the 1950s. It appears that 'arithmetic' was considered most appropriate for the mainly working-class pupils in secondary modern schools, whereas the predominantly middle-class pupils in grammar schools studied 'mathematics'.

The 1960s saw the introduction of comprehensive schools, and the mathematical needs of all pupils were gradually reassessed. Recommendations that *mathematics* rather than simply arithmetic should become part of the curriculum for *all* pupils up to the age of 16 followed from a report that there were substantial differences in the amount of time allocated to the study of mathematics between British pupils taking science-based courses – who spent approximately 22% of their total time studying mathematics – and those taking humanities-based courses – who spent only 10% of their time on mathematics (see Griffiths and Howson 1974). Moreover, though the mathematical attainment of science students, who were mostly male, compared favourably with the rest of Europe, that of arts and humanities students, who tended to be female, had fallen behind.

Subsequently there have been major changes in the organisation of mathematics education, both in terms of the time traditionally allocated to mathematics and in the content and method of teaching. Again, much of the discussion about the suitability of including certain mathematical topics has been related to pupils' future lives and occupations.

Activity 1.6

List the reasons why you think pupils should study mathematics. Do you agree that the mathematics curriculum should be related to the economic requirements of society? Do different groups of pupils have different mathematical requirements?

Comment

School mathematics has been increasingly justified on the basis of the economic needs of society. The Dainton Committee (1968) in particular was concerned about the ways in which mathematics should help to prepare pupils for future employment (as was the Cockcroft Report 1982). Dainton listed the following uses for the study of mathematics:

- as a means of communicating quantifiable ideas;
- as a training for discipline of thought and for logical reasoning;
- as a tool in activities arising from the developing needs of engineering, technology, science, organisation, economics, sociology etc.;
- as a study in itself, where development of new techniques and concepts can have economic consequences akin to those flowing from scientific research and development.

Bearing in mind that Dainton was concerned about the vocational uses of mathematics, can you detect a specific orientation towards either girls or boys, towards either more able or less able pupils, or towards groupings of pupils according to race or class background in any of the reasons listed above?

What can be done?

This section has dealt mainly with changes to the mathematics curriculum over the last century and so perhaps does not address directly the immediate needs of today's classroom. Nevertheless it is hoped that this brief look at the way school mathematics has changed over the years will enable you to reflect on the content of the mathematics curriculum in your own school. You and your colleagues might like to consider:

(1) the reasons why certain topics (and not others) are included in mathematics syllabuses (see also section 1.4);
(2) the hidden messages that the inclusion and treatment of certain topics might give to different groups of pupils (see also section 1.4);
(3) whether textbooks and resources used in the mathematics department take account of the experiences and future lives of *all* the pupils in the school, and not just those who are male, white and middle class, as often seems to be the case (see chapter 4).

1.3 Women's contributions to mathematics and the implications for girls today

Recently a university lecturer, during a course on modern algebra which dealt mainly with rings and modules, told his students a little of the history of the topic. He mentioned that the most significant developments had been made by Emmy Noether and told them a little about her life. The reaction from his class of post-graduate students, the majority of whom were mathematics teachers, astonished him. The women students in particular were amazed and delighted to hear that such a major contribution to mathematics had been made by a woman. 'I did not know that there had ever been any women mathematicians,' said one student.
(Barnes, Plaister and Thomas 1984)

It has been claimed that one of the most important factors discouraging girls from the study of mathematics is the way in which society has viewed it as a male domain. In fact, in the past women have made a number of contributions to the study of mathematics, though it has been argued that their work has been systematically devalued because of their gender (see Spender 1982b; Leder 1984). These include the following Europeans:

26

- *Hypatia* (Greek 375–415). She occupied the Chair of Platonic Philosophy at Alexandria and taught geometry, astronomy and the new science algebra. She is credited with several inventions for use in everyday life – an apparatus for distilling water, an instrument for measuring the specific gravity of water, an astrolabe and a planisphere.
- *Gabrielle du Chatelet* (French 1706–46). She was the first person to translate Isaac Newton's *Principia Mathematica* into French; her long essay, 'Les Institutions de Physiques', written in 1740, introduced into France the ideas of Leibniz. She lived for many years with Voltaire, working on the nature of fire with him. In a study submitted to the French Academy of Sciences in 1738, she anticipated later research by maintaining that heat and light have the same cause or are both modes of motion.
- *Maria Gaetana Agnesi* (Italian 1718–99). At the age of 9 she spoke in Latin for an hour to a learned assembly on the rights of women to study science. Her major work on differential and integral calculus, *Le Instituzioni Analitiche*, was translated into several languages and she was offered, but did not accept, the Chair of Higher Mathematics at Bologna.
- *Sophie Germain* (French 1776–1831). Mathematical physicist. She won the prize offered by the French Academy of Science to the one who could 'give the mathematical theory of the vibration of elastic surfaces and compare it with the results of experiment'. Gauss recommended that she be given an honorary doctorate by Göttingen University, but she died before it could be awarded.
- *Mary Somerville* (British 1780–1872). She translated Laplace's book on the mechanism of the heavens into English, and, together with Caroline Herschel, was elected an honorary member of the Royal Astronomical Society. She wrote *Physical Geography*, which was greatly admired by Humboldt, and several monographs on mathematical subjects, e.g. curves and surfaces of higher orders. In recognition of her services to science she was granted a pension of £300 per year by the government. Her signature was the first on John Stuart Mill's petition to Parliament for women's suffrage.
- *Ada Byron Lovelace* (British 1815–52). Daughter of Lord Byron, the poet. Soon after Ada was born, her parents' marriage broke up, leaving Lady Byron very bitter about her marriage. Lady Byron was very concerned about Ada's education, particularly in mathematics. When she was 13 or 14 Ada decided she wanted to become a famous mathematician. She met Charles Babbage, the inventor of the computer, who suggested that she publish a translation of the work of an Italian engineer and add her own notes about computers as well; she also described, elsewhere, how Babbage planned to use punched cards to instruct his analytical engine (computer) to carry out complicated computations.
- *Sonya Kovalevsky* (Russian 1850–91). She studied mathematics at Heidelberg and won a degree *in absentia* from Göttingen for a brilliant thesis on partial differential equations. She won the French Academy Prix Bordin for her work on the rotation of a solid body about a fixed point. She was appointed Professor of Higher Mathematics in the University of Stockholm and, shortly before her death, was elected to the St Petersburg Academy of Science.
- *Emmy Noether* (German 1882–1935). She made a significant contribution to modern algebra with her studies on abstract rings and ideal theory. As a Jew, an intellectual and a liberal, she was dismissed from her appointment at the University of Göttingen in 1933 during the Nazi rise to power; she became a professor at Bryn Mawr College in the USA.
- *Grace Hopper* (American 1906–). Computer pioneer. Worked for the UNIVAC division of Sperry Rand on the first compiler – the translator between humans and computers. She developed the concept of automatic programming which led to COBOL (Common Business Oriented Language). (Smail 1984)

These women faced enormous obstacles in order to pursue their studies.

Perhaps this provides not only an indication of their commitment to the study of mathematics, but also an explanation for the relatively few women mathematicians to have gained public recognition.

Activity 1.7

The extract from the *Edinburgh Review* in figure 1.7 summarises Mary Somerville's experiences in studying mathematics. Jot down the types of barriers she faced and also any support she received.

Figure 1.7   Mary Somerville's experiences (*Edinburgh Review* 1934, p. 155)

Her parents opposed her mathematical studies but she was helped by her brother's tutor. Marriage to her cousin Samuel Grey gave her some freedom though her situation entirely changed when on his death after only three years of married life, she was left financially independent. She began to study mathematics and astronomy in earnest and her second husband supported her in her studies. She made contact with a number of European mathematicians including Pierre Laplace, whose *Mecanique Celeste* she translated into English. A contemporary review described the translated works as giving a 'condensed and perspicuous view of the general principles and leading facts of physical sciences, embracing almost all modern discoveries which have not yet found their way into our elementary work'.

Comment

Barnes, Plaister and Thomas (1984) list the following barriers to women's achievement in mathematics in the past, many of which were applicable to Mary Somerville:

- *Academic education was not generally available to women.* Women who openly expressed a wish to undertake serious study invariably had to face accusations of eccentricity and doubts expressed about their womanliness and normality, particularly in relation to child bearing. In fact Mary Somerville received no formal mathematics education.
- *Opposition from families.* Some parents used extreme measures to prevent their daughters from studying. For instance, Sophie Germain's parents took away not only heating and lighting from her room but also her clothes. Mary Somerville faced similar opposition from her parents.
- *Until comparatively recently, universities excluded women both as students and as lecturers.* In order to pursue their studies, women mathematicians had to circumvent the system. For instance, Sophie Germain borrowed lecture notes from friends and Sonya Kovalevsky entered into a marriage of convenience in order to study in Germany.
- *Most women had little chance of access to mathematical education, ideas or books.* It is no coincidence that many successful women mathematicians had husbands, brothers, fathers or mentors who were themselves mathematicians. (Mary Somerville received such support from her brother's tutor, and also from European mathematicians.)
- *For married women, children presented further problems.* Women from aristocratic families could leave the care of their children and households to servants. Others, like Mary Somerville, had to fit their mathematical studies around their domestic duties.

- *Even established mathematicians lacked belief in their ability to do original and creative work.* A recurrent feature of women mathematicians' work is the translation and expounding of other peoples' contributions. For example, Mary Somerville translated Laplace's work into English, Gabrielle du Chatelet translated Newton's *Principia Mathematica* into French, and Mary Agnesi wrote a treatise on calculus which brought together the work of Newton, Leibniz and many others. Mary Somerville, one of the greatest of all mathematicians, showed clearly the misgiving that academic women had about their own performance when she said: 'I have perseverance and intelligence but not genius. That spark from heaven is not granted to the [female] sex . . .'

Activity 1.8

Whilst many of these past barriers to women's achievement in mathematics have been removed, can you identify any that exist today?

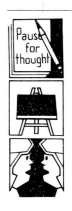

  (i) Compare your experience with that of Mary Somerville; jot down the obstacles that you had to overcome and the support that you received.
 (ii) Ask the pupils in your tutorial group or one of your classes to do the same and compare the results.

Comment

In the UK there are now few *formal* barriers to the study of mathematics by girls and women – mathematics is one of the compulsory subjects that all pupils take up to 16 and there are few overt restrictions to advanced levels of study. Yet many would argue that remnants of past barriers to female mathematical achievement remain today (see, for example, Isaacson (1986) in *Girls into maths can go*). Mathematics still has an 'unfeminine' or 'blue-stocking' image; in fact, several women teachers admitted in the workshops that in social situations they attempt to disguise the fact that they teach mathematics. Also, women and girls continue to devalue their ability in mathematics and other scientific subjects; they are far too easily convinced of their inability to tackle what they see as 'difficult' subjects (see chapters 2 and 3). Even today many women and girls face opposition from their families, or have to fit their mathematical studies around domestic duties (see Leder (1984) in *Girls into maths can go*).

You might also like to discuss with your pupils whether it is particularly difficult for some groups of girls and boys to study mathematics because of their cultural or class background.

What can be done?

The experiences of women mathematicians in the past can be used to illustrate the problems that girls and women teachers (even perhaps men teachers and boys) may be facing today. This information can be of direct and practical use in the classroom.

(1) It provides all pupils with a social context for their mathematical learning and so could help to 'humanise' a subject often regarded as rather 'cold'.
(2) It provides girls with examples of successful role-models, and boys with evidence of contributions that women have made in the past. All pupils could thus be encouraged to widen their often very narrow assumptions about the possible achievements of girls and women.
(3) In this section we have concentrated only on the contributions of

pre-twentieth-century European women because this information was readily available. But there are a number of well-known twentieth-century female mathematicians (e.g. Dame Mary Cartwright, Olga Todd, Alice Roth, Grace Hopper) whose contributions and experiences might be worth investigating. Also, mathematics as we know it today is derived from many different cultures, so you might like to investigate the contributions of women and men from, for example, Asia and Africa.

(4) Finally, the appendix 'Herstory of mathematics' in *Girls into maths can go* (Burton 1986) might be used in the form of case studies as a basis for discussion about gender and race stereotyping and the restrictions it imposes on the motivation and life-chances of individuals. It might provide a useful link between social studies and mathematics (perhaps you could liaise with social studies teachers).

1.4 The mathematics curriculum today

In section 1.2 we mentioned that in recent years there have been a number of major changes to the content and method of teaching mathematics. We now examine these changes and consider the implications for girls.

Activity 1.9

Figure 1.8 (adapted from GCE O-level and CSE syllabuses) provides some indication of the mathematics content covered at secondary level (11–16) in 1955 and in 1985.

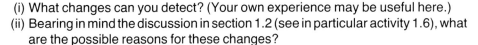

(i) What changes can you detect? (Your own experience may be useful here.)
(ii) Bearing in mind the discussion in section 1.2 (see in particular activity 1.6), what are the possible reasons for these changes?

Comment

Inevitably, individuals and groups have different opinions about the process of change in their schools. Some teachers feel that in many ways mathematics has not changed very much at all, notwithstanding the fact that in 1955 it tended to be regarded only as a grammar-school subject whereas today it is studied by pupils of all abilities. It is argued that many of the topics still seem to be difficult, dry, unappealing and of little relevance to the future lives of many pupils.

Yet there have been a number of changes, the most obvious being in geometry and arithmetic. For example, in 1955 there was considerable emphasis on geometric constructions and on the proof and application of a seemingly endless number of theorems in the tradition of Euclid (in this sense geometry had not changed for about 2000 years!). Today there is a wider interpretation of the word 'geometry'; children study tessellations, transformations and vectors in addition to a less rigid treatment of the properties of lines and shapes. Also, the advent of the calculator and the increased use of the metric system have led to major changes in arithmetic; today there is less emphasis on long multiplication, long division and the use of mathematical tables, and more work on estimation and approximation. Several new topics have been introduced – such as set theory, calculus, matrix algebra, probability and statistics. You may have detected other changes as well.

There are probably a number of reasons for such changes, many of which are

Figure 1.8   Secondary level mathematics (11–16) 1955 and 1985

1955	1985
Arithmetic Numbers; fractions (including vulgar fractions); the four rules; measurement of length, area, volume, capacity, weight, time in English and metric units; mensuration; averages; ratio; proportion; percentages; simple and compound interest; practical applications of arithmetic; use of logarithms to base 10.	*Arithmetic* Numbers; decimals; simple fractions; the four rules; use of SI units; ratio; proportion; percentages; practical applications of arithmetic; estimation and approximation; four-figure tables are provided, and the use of electronic calculators and slide rules is permitted.
Algebra Fundamental processes; symbolic expression of general results; interpretation and evaluation of formulas; changing the subject of a formula; direct and indirect variation; factor and remainder theorems; indirect factorisation of various types; equations of 1st and 2nd degree; simultaneous equations of 1st degree and two simultaneous equations of which one is 1st degree and one is 2nd degree; easy examples in fractions; graphs of simple algebraic functions with applications; indices (including negative and fractional); arithmetic and geometric series; application of algebra to the solution of problems.	*Algebra* Algebraic manipulation and evaluation – including ability to change the subject of a formula; use and notation of set theory; factor and remainder theorems; factorisation of various types; solution of linear and quadratic equations; solutions of simple inequalities (linear and quadratic); geometric representation of linear inequalities in two variables; arithmetic and geometric series; use of binomial theorem, manipulation of scalar and vector quantities; use and properties of indices and logarithms; graphs of polynomials and rational functions with linear denominators; simple vector and matrix algebra; application of algebra to the solution of problems.
Geometry Construction of perpendicular to straight lines; construction of equal angles and angles of 60°, 45°, 30°; division of a straight line into a number of equal parts or two or more parts of given proportion by construction; construction of triangle equal in area to a given polygon; construction of tangents, circumscribed, inscribed and escribed triangles; construction of equiangular triangles to given triangle; proof and application of theorems on straight lines, triangles, rectilinear figures, circles; similar triangles; areas; loci.	*Geometry* Use of vectors to establish properties of geometric figures; collinearity; concurrency; condition for two lines to be perpendicular or parallel; simple properties of the circle; tessellations, rotations, reflections, other transformations and their representation using matrices; point dividing a line in a given ratio.
Trigonometry Sine; cosine; tangent; use of tables and application to practical problems in two and three dimensions; bearings, latitude and longitude; sine rule; cosine rule; secants; cosecants; cotangents.	*Trigonometry* Sine; cosine; tangent; application to practical problems in two and three dimensions; use of sine and cosine addition formulas.
	Calculus Differentiation and integration of powers of x; applications to simple kinematics and evaluation of areas and volumes; maxima and minima; equations of tangents.
	Probability and statistics Simple probabilities – definitions and evaluation; measures of average; measures of spread; ways of representing data diagrammatically – bar charts, pie charts, frequency curves.

economic and social. For example, the introduction of probability and statistics is partly due to the fact that statistics are increasingly used by the media, advertising and industry (i.e. these topics now appear because they are relevant to pupils' future lives). The increasing emphasis on estimation and approximation can be similarly explained. The introduction of topics such as calculus and vectors is partly explained by the changing needs of other scientific and technical subjects. On the other hand, it is possible to detect some purely academic changes. For example, changes in notation and terminology (rightly or wrongly) may simply reflect advances in higher-level mathematics. Some of the changes have come about to make the subject more accessible to children of all abilities.

Do these changes have any implications for girls in their study of secondary-level maths? It might be argued that most mathematical topics have no specific internal bias and that the bias usually lies in the context in which they are presented and learned. In this sense, many topics and applications have always had particular orientation towards the experiences of men and boys (for example, topics in trigonometry and kinematics, applications to building, navigation, stocks and shares etc.). Figure 1.8 suggests that little has changed in this direction; a large number of topics which appear in the 1985 maths curriculum do so because of their importance in other traditional maths-related subjects which, as we showed in section 1.1, are studied by more boys than girls. Looking down the list, there are very few topics which are directly relevant to the subjects favoured by girls!

Nevertheless, there is some evidence to suggest that direct curricular intervention *can* have an influence on overall mathematical performance. In a study of 200 children over one year, F. Howard Thomas (1983) found that the general mathematical ability of pupils taught transformation geometry showed greater improvement than their peers using traditional schemes of work and, moreover, that the improvement was more marked for girls. This study involved four sets of learners: girls following a traditional syllabus (TG), boys following a traditional syllabus (TB), girls following a new syllabus (MG) and boys following a new syllabus (MB). Thomas compared the average scores of these four sets of learners on standard tests at the beginning and at the end of the year. Figure 1.9 indicates his results.

Figure 1.9 Comparison of initial and final average scores of Thomas' four sets of learners. (*Note.* The scores were standardised in such a way that on both the initial and final tests the boys following the traditional syllabus (TB) had an average score of 100.)

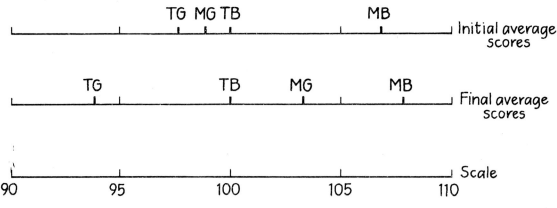

32

This study suggests that changes in the actual topics covered could have a beneficial effect on girls' attainment. On the other hand, it is not possible to conclude from Thomas' research that girls' performance in maths examinations is necessarily improved. For instance, Robert Wood (1977) found when examining O-level syllabuses in 1973 and 1974 that girls simply tended to avoid certain topics – particularly those in probability and geometry! Shiam Sharma and Roland Meighan (1980) came to a similar conclusion; although girls do better in 'modern' maths, they tend to avoid three-dimensional problems in trigonometry and geometry.

Of course, in addition to changes in the content of the maths curriculum, there have also been a number of significant changes in the way that mathematics is taught and learned. There is now considerable emphasis on process rather than just the end results. Current thinking in mathematics education can be summarised by figure 1.10, which lists the aims for all pupils following courses leading to GCSE examinations (the list is taken from the *National criteria for mathematics* (GCE and CSE Boards' Joint Council 1985). These aims (or adapted versions of them) are to be adopted by all secondary schools in the next few years.

Figure 1.10   Aims of GCSE maths courses

All courses should enable pupils to:
2.1 develop their mathematical knowledge and oral, written and practical skills in a manner which encourages confidence;
2.2 read mathematics, and write and talk about the subject in a variety of ways;
2.3 develop a feel for number, carry out calculations and understand the significance of the results obtained;
2.4 apply mathematics in everyday situations and develop an understanding of the part which mathematics plays in the world around them;
2.5 solve problems, present the solutions clearly, check and interpret the results;
2.6 develop an understanding of mathematical principles;
2.7 recognise when and how a situation may be represented mathematically, identify and interpret relevant factors and, where necessary, select an appropriate mathematical method to solve the problem;
2.8 use mathematics as a means of communication with emphasis on the use of clear expression;
2.9 develop an ability to apply mathematics in other subjects, particularly science and technology;
2.10 develop the abilities to reason logically, to classify, to generalise and to prove;
2.11 appreciate patterns and relationships in mathematics;
2.12 produce and appreciate imaginative and creative work arising from the mathematical ideas;
2.13 develop their mathematical abilities by considering problems and conducting individual and co-operative enquiry and experiment, including extended pieces of work of a practical and investigative kind;
2.14 appreciate the interdependence of different branches of mathematics;
2.15 acquire a foundation appropriate to their further study of mathematics and of other disciplines.

Activity 1.10

(i) From your own experience, what types of changes do these aims represent?

(ii) What (if any) are the implications for girls in these latest developments in mathematics education?

Comment

The aims listed in figure 1.10 confirm that the structure of school mathematics has changed from being content- and topic-based towards an emphasis on process. Part of this change has meant an increased focus on mathematical relevance, problem solving and investigation, because more has been learnt about the difficulties in understanding and applying mathematical abstractions. This might have particular importance for girls, since they are rather less likely to be motivated by a study of mathematics for its own sake and more likely to be influenced by the relevance of the material to their own lives and their future careers (see sections 2.1 and 2.4).

There is more emphasis on communication skills and on the interpretation of results. This could also be to the benefit of girls, since they tend to perform better than boys in the subjects which have traditionally involved such skills (such as English, languages, history). Also, traditional teaching methods do not necessarily help pupils to develop the skills needed to achieve the aims listed here; it could be that the most appropriate teaching and learning approaches are more suited to girls' needs (and also to the needs of those boys who are disadvantaged in some way).

What can be done?

There are a number of ways in which teachers can attempt to counteract bias within the mathematics curriculum in their schools.

(1) Try to ensure that maths topics are presented in a challenging way for all pupils and that they are seen as directly relevant to a wide range of subjects, i.e. to those which are popular with girls as well as those favoured by boys (see section 1.1). Make links with biology, social studies, home economics and art as well as with physics and geography.

(2) Try to avoid any gender and race bias in 'real life' applications. That is not to say that all teaching should be entirely abstract. Rather, draw on (say) knitting examples as well as cricket scores to illustrate concepts and use domestic as well as commercial contexts (but not always in stereotyped situations) to demonstrate applications. Role reversals could be applied *without comment* to show, for instance, women working in the City or in the factory and men occupied in domestic tasks or in the garden, and illustrations should be provided which depict our multi-cultural society. See chapter 4 for a fuller discussion of these issues.

(3) Think about the *way* that you teach; could you encourage more use of the types of skills where girls excel (e.g. communication skills)? One suggestion is that you might encourage pupils to deal with certain unfamiliar mathematical topics through visualisation and imagination. For instance, in dealing with velocity, it might be an idea to ask pupils to form a mental image of riding a bike and to talk through their imagined experience, in order to understand better the mathematical concept. The use of imagery could be particularly appropriate when teaching topics in geometry. For some ideas on this see Kent and Hedger (1980), Shuller (1983) and ATM (1982).

1.5 Summary and reflection

This chapter has provided a general overview of the development of school mathematics over the past century, and its particular implications for girls. It included a brief discussion about women mathematicians and their struggles to continue their studies, in order to highlight many of the problems that still exist today. We showed how the mathematics taught in other school subjects can affect girls' mathematical performance. Finally we indicated recent and proposed changes to the mathematics curriculum and considered the extent to which these might help girls.

Activity 1.11

On reflection, which of the suggested strategies for change will have the most likelihood of adoption by the staff in your school? Jot down in the table below the aspects of the problems discussed in this chapter which are most relevant to your pupils or to your school. Beside each of these, note the strategies that you feel you can most effectively introduce to improve the situation, either now or in the future.

Aspect of the problem	Strategies for tackling it

Further reading

- Articles by G. Leder, Z. Isaacson and the appendix in *Girls into maths can go*
- T. Perl, *Math equals: biographies of women mathematicians and related activities* (Addison Wesley)
- L. M. Osen, *Women in mathematics* (MIT)
- *National criteria for mathematics* (HMSO)
- *Mathematics from 5 to 16 (Curriculum matters 3: an HMI series)* (HMSO)

2 Feelings, attitudes and expectations

Some boys and girls do really bad in maths. It is not because they are stupid though . . . I myself am very bad at maths . . . you know [that people are bad at maths] by their attitude to maths . . . boys and girls panic or run out of time.

Even if you are not very good at maths I think it is best to try and do things you don't understand as best you can.

I don't like maths. I don't suppose anyone feels the way that I do.

(Fifteen-year-old girls talking about mathematics in Joffe and Foxman 1986)

'Strong, usually negative feelings are often engendered by the mere mention of the word "mathematics",' reported the Assessment and Performance Unit (APU 1981) in its research into pupil attitudes. Whilst many boys probably have similar feelings to girls, the APU found that girls were likely to express greater uncertainty about their mathematical abilities and performance whereas boys had greater expectations of success. The opening quotes typify the feelings expressed by many more girls than boys:

- feelings of dislike and irrelevance;
- feelings of panic, anxiety and fear;
- feelings of isolation and lack of confidence;
- feelings of inadequacy and stupidity.

Activity 2.1

Can you remember ever having similar feelings to these, either when learning mathematics yourself or during your years of teaching the subject? Jot down the situations that gave rise to such feelings and also your reactions.

Comment

Whatever negative feelings you may recall, it is likely that they were caused by one of the following:

- some sort of failure (failure to understand something, or not doing something as successfully as you might have liked);
- some sort of pressure (pressure from colleagues and pupils, or pressure due to lack of time);
- some observation of irrelevance (the irrelevance of a particular topic, or the seeming irrelevance of a piece of work or a way of thinking simply because that was not the *expected* response).

Did you react by 'switching off'? Did your mind go blank? Did you experience 'sweaty hands' or 'butterflies in the tummy'?

For many people, their first experience of the feelings mentioned above – and the situations that give rise to them – is in school. The fact that so many girls associate such feelings with mathematics in particular suggests that the reasons why girls under-achieve in this subject and 'opt out' are extremely complex. Feelings of insecurity, anxiety and lack of confidence cannot be explained simply in terms of the 'maleness' of the subject as discussed in chapter 1. Activity 2.1 indicates that pupils' feelings towards mathematics also owe much to the attitudes of the pupils themselves, heavily influenced by their peers, teachers, parents and society in general. It is this aspect of the problem that we shall consider in this chapter.

We begin by investigating in more detail the attitudes of pupils, considering in what ways these might influence mathematical performance. We then examine the attitudes of teachers (in section 2.2) and of parents (in section 2.3) and discuss the extent to which these help to shape and reinforce the views of children and young people. Finally we investigate the problems faced by pupils in relation to employers and employment. In each section we suggest ways in which you might try to discourage stereotyped views in order to help all pupils develop more positive feelings towards mathematics.

2.1 Pupils' attitudes

During the development of this pack teachers at workshops suggested that girls tend to have more negative attitudes towards mathematics than boys because:

- girls see mathematics as more useful to boys;
- being good at maths involves risks and girls are not encouraged to take risks;
- being good at maths can conflict with adolescent conformity to sex-role stereotypes – the images of 'macho' boys and 'feminine' girls are at variance with that of the good mathematician;
- girls place more importance on how much they like a subject and they tend to 'dislike' maths more than boys;
- girls feel more anxious in mathematics lessons because of their attitudes towards success and failure.

Do you agree with these very general observations? Do you think that different groups of girls and boys react differently – for example, can you identify any differences in the attitudes of pupils from different cultural and class backgrounds?

Activity 2.2

What exactly are the attitudes of your pupils? Prepare enough copies of the questionnaire in figure 2.1 for all the pupils in your tutorial group or in one of your maths classes.

 (i) Ask pupils to complete the questionnaire.
(ii) For each question and for each sex count:

   (a) the number of ticks;
   (b) the number of crosses;
   (c) the number of blanks.

(iii) Compare the responses of girls and boys.

Notes

(a) It should take about ten minutes for pupils to complete the questionnaire. Tell them it is intended to be anonymous. The only identification required is that they should indicate whether they are male or female.

(b) You may like to adapt the questionnaire slightly in order to investigate other differences in attitude. For example, you may like to compare the responses of pupils of different ages. You might also like to compare the responses of pupils from different cultural and class backgrounds.

(c) You might like to ask the pupils to help in analysing the data. One teacher who tried this reported that it led to some interesting work on ways of displaying and analysing data.

Comment

Keep your results handy; they will be referred to throughout this chapter and in chapter 3.

American researchers were the first to explore in pupils the relationships between attitudes and achievement. Not surprisingly, they soon discovered that positive attitudes towards mathematics were linked to self-confidence and to achievement. In the rest of this section we explore these links and consider the particular implications for girls.

Interest: like/dislike of the subject

Activity 2.3

(i) Look back at the responses to statements 8, 13 and 14 on the questionnaire. What percentages of girls and boys (a) agreed with, (b) disagreed with, or (c) were neutral about these statements?
(ii) The responses to these statements provide some indication of pupils' interest in mathematics since usually people enjoy doing something that they find interesting. Do any distinct patterns emerge in your results: overall, or with particular reference to girls and boys, or with reference to pupils from different age-groups or from different cultural and class backgrounds? Are there any surprises for you? You might like to discuss your findings with your pupils.

Comment

In 1981 the APU found that girls enjoyed mathematics less than boys and, in particular, that more boys than girls rated geometric and measurement topics as interesting (and useful) (APU 1981). As we discussed in chapter 1, girls may well dislike mathematics more than boys do because they view it as a 'male subject'. Researchers in America and in Britain have found that girls tend to see mathematics and the sciences as 'hard, intellect-based, complex and masculine' and that these views seem to sharpen at about the age of thirteen or fourteen plus (Fennema 1980); so it is likely that the older the pupils, the greater the differences in response to these statements. Do your results support this hypothesis?

Of course, it is not necessary to be good at something to enjoy doing it, but there does seem to be evidence of a relationship in the other direction – particularly for girls. For example, the APU reported:

> It appears that girls who find individual topics interesting are more apt to find them easy and to view mathematics as a less difficult subject. This relationship is less strong in the case of boys. (APU 1981)

Although, in general terms, the *direct* link between enjoyment of mathematics and performance in written tests is weak, Lynn Joffe and Derek Foxman in *Girls into maths can go* point out that:

> This does not mean that enjoyment is not important, since it could affect how much mathematics pupils choose to study and . . . enjoyment of the subject should be an educational goal itself. (Joffe and Foxman 1986)

Figure 2.1 Your feelings towards mathematics (adapted from Barnes, Plaister and Thomas 1984)

Read the following statements carefully, and tick those which describe the way you feel, put a cross against those you disagree with and leave the boxes blank if you have no particular feelings either way.

☐ 1. I prefer to work on my own.
☐ 2. You have to be brainy to do well at maths.
☐ 3. I enjoy working with friends.
☐ 4. I usually understand a new idea in maths quickly.
☐ 5. I usually get most of my maths right.
☐ 6. I usually feel confident about maths tests.
☐ 7. Boys generally do not like it if girls beat them at maths.
☐ 8. I really enjoy solving maths problems.
☐ 9. Knowing maths will help me get a job.
☐ 10. I always feel nervous when I look at a maths problem.
☐ 11. I am lucky when I do well on a maths test.
☐ 12. I feel nervous when I am asked questions in maths.
☐ 13. I am disappointed when I miss a maths lesson.
☐ 14. I cannot understand how anyone could enjoy maths.
☐ 15. If I work carefully I find maths easy.
☐ 16. If I do well at maths some people make fun of me.
☐ 17. I do not see the point of most of the maths we do.
☐ 18. I think my maths teacher thinks I'm stupid
☐ 19. I think my maths teacher enjoys teaching me.

In the following statements, tick the explanation which is most important for you:

20. If I do well in maths it is usually because:
   ☐ I'm naturally good at it.
   ☐ I work very hard.
   ☐ I have a good teacher.
   ☐ The work is very easy.

21. If I do badly in maths it is usually because:
   ☐ I'm always hopeless at it.
   ☐ I did not try hard enough.
   ☐ I was unlucky.
   ☐ The work was too hard.

22. When I need help with maths at home, I usually ask _____

	Girl	Boy
23. Please tick	☐	☐

Activity 2.4

(i) Look back at the responses to statements 2, 7 and 16 on the questionnaire. What percentages of girls and boys (a) agreed with, (b) disagreed with, or (c) were neutral about these statements?

(ii) The responses to these statements provide some indication of pupils' perceptions of each other. Do any distinct patterns emerge in your results: overall, or with particular reference to girls and boys, or with reference to pupils from different age-groups or from different cultural and class backgrounds? Are there any surprises for you? You might like to discuss your findings with your pupils.

Comment

During the development of this pack teachers found that more girls than boys agreed with these statements. The age of the pupils could also be an important factor, for two reasons. First, according to a small study conducted by Gilah Leder (1980), it seems that at adolescence many boys value success most highly while many girls consider popularity with peers most important; and that girls are far more likely to have high aspirations if they have friends who have similarly high expectations. Second, boys increasingly 'annexe' mathematics as a 'male domain' as they proceed up the school (Fennema and Sherman 1976), and the strong peer-group influence during adolescence leads to an increasing acceptance of maths as an 'unfeminine' subject by both girls and boys (although patterns are likely to vary between groups of pupils from different cultural and class backgrounds).

As part of Gilah Leder's study (1980), pupils were given the following story-line with the Anne cue being given to the girls and the John cue to the boys.

Anne (John) came top of her (his) mathematics class. Describe Anne (John).

Leder's results may be summarised in her own words:

While it is difficult, if not dangerous, to draw any firm conclusions from such anecdotal data derived from a small sample, it appeared that the boys interviewed were less ambivalent in their responses to the successful male cue figure than were the girls in their responses to the successful female figure.

Activity 2.5

(i) Use the Anne (John) story with some of your pupils. For each sex, give half the pupils the Anne cue and half the John cue.

(ii) Count up the number of times various words or phrases are mentioned. Do any patterns emerge?

Comment

This activity was carried out by several teachers during the development of this pack. They were surprised at how unflattering children can be about anyone who is good at maths. Two typical responses were:

Anne is a hard-working girl who finds maths rather easy. She has ginger hair with glasses and buck-teeth. Her hair is greasy and she has a spotty forehead. She has lots of freckles. She always works quietly and does her homework. She wears pebble glasses.

John is fairly brainy, works hard, also concentrates on his work. I think John is a boffin. I think John would be a puny skinny kid with National Health glasses. Also he does his top button on his shirt up with his tie done right up to his neck. He would wear stupid horrible shoes as well.

Figure 2.2 Comments about Anne and John

	Favourable about Anne	Unfavourable about Anne	Favourable about John	Unfavourable about John	Total
Girls	3	6	4	5	18
Boys	2	7	2	7	18
Total	5	13	6	12	36

Figure 2.2 indicates how one teacher classified the results of his class. The kinds of comments which were made frequently by this class were:

	Anne	John
Freckles/glasses/old-fashioned hair styles	8	9
Neat/tidy	5	1
Big headed/bossy	5	4
Well liked/makes friends easily	3	4
Not well liked	2	2

Comments on physical appearance all tended to be derogatory but particularly of Anne who was described as 'flat chested' with 'straight hair' and 'goofy teeth'.

The results suggest few differences in the opinions of the girls and the boys other than that both disliked anyone good at mathematics, although the girls did seem to think more favourably of a boy good at maths, than the boys. Do your results fit this pattern?

Again, whilst there is no evidence of a direct relationship between pupils' perceptions of each other and girls' under-achievement in mathematics, there must surely be some indirect link. The popular view seems to be that being good at mathematics and being attractive and feminine are incompatible and so once more mathematics is perceived as somehow inappropriate and masculine. Adolescent girls are therefore likely to have less motivation to succeed in mathematics, with inevitable consequences. (*Note:* You may wish to explore here the views of pupils from different cultural and class backgrounds. It could be the case for example that some black boys are also deterred by the image of successful white mathematicians.)

Confidence and anxiety

Activity 2.6

(i) Look back at the responses to statements 4, 5, 6, 10 and 12 on the questionnaire. What percentages of girls and boys (a) agreed with, (b) disagreed with, or (c) were neutral about these statements?

(ii) The responses to statements 4, 5 and 6 provide some indication of pupils' confidence in mathematics; those to statements 10 and 12 indicate levels of anxiety. Do any distinct patterns emerge in your results: overall, or with particular reference to girls and boys, or with reference to pupils from different age-groups or from different cultural and class backgrounds? Are there any surprises for you? You might like to discuss your findings with your pupils.

Comment

Research has shown that in general girls exhibit less confidence than boys in their mathematical ability, and that girls' confidence decreases whereas boys' confidence increases as they progress through school. Do your results fit this pattern?

Part of a five-year investigation carried out in Sheffield schools (Eddowes 1983), which involved a large sample of twelve-year-old pupils who were examined at the beginning and the end of one school year in arithmetic and problem solving, included several attitude surveys. In the attitude survey at the beginning of the year the researchers found that gender differences were small overall compared with general school differences, though girls showed less confidence in their ability to cope with mathematics and thought the subject difficult. By the end of the year, however, the overall attitude mean score was significantly lower for girls. In particular, measures of anxiety and lack of confidence were increased, though this was also true of the boys but to a lesser extent.

This general trend may not be exhibited by *all* girls and boys however. For example, the Swann Report (1985) suggests that many black boys also experience loss in confidence as they progress through the school.

As has already been mentioned, the APU (1981), in its survey of fifteen-year-olds, found girls to be more uncertain than boys about their mathematical performance. The study also found that

> Boys overrate their performance in mathematics in relation to written test results; they do not do as well as they expect to. Girls underrate their performance and do better on tests than they expect.

So it seems that not only are girls held back in mathematics by lack of confidence and misplaced modesty, but where they do succeed, they do not 'believe their eyes' and continue to belittle their achievements. In this sense their experiences are similar to those of eminent women mathematicians of the past, as discussed in section 1.3. Increasing all pupils' confidence could well help to improve their performance in mathematics, but in order to do this it is necessary to be able to recognise behaviour which typifies confidence or lack of confidence.

Activity 2.7

Jot down a few characteristics which you feel are indicative of confidence or lack of confidence.

Comment

Figure 2.3 indicates the types of behaviour identified by teachers at workshops. Are your views similar?

Figure 2.3 Typical characteristics of confidence and lack of confidence

Pupils seem confident when they	Pupils indicate lack of confidence when they
■ work independently; ■ share their ideas; ■ show interest and want to go further; ■ exhibit sustained interest; ■ show creativity in tackling new work and ideas.	■ repeatedly ask what to do next; ■ are disruptive or withdrawn; ■ tend to ask trivial or irrelevant questions; ■ copy out questions to save thinking; ■ show reluctance or distress when asked to tackle new work; ■ appear to enjoy only repetitive work.

Activity 2.8

What evidence is there of a relationship between confidence and ability in mathematics?

(i) Make a list of the five most able pupils (girls or boys) and the five least able pupils in one of your maths classes.
(ii) Jot down which of the characteristics indicated in figure 2.3 might be associated with each of these pupils.
(iii) To what extent is the relationship between confidence and ability established in your small sample?

Comment

Do the able pupils tend to display the behavioural characteristics of confidence and do the weak pupils display those of lack of confidence? Are there more girls than boys (or vice versa) in your lists of able and weak pupils and if so what does this suggest? Can you identify any patterns for pupils from different cultural and class backgrounds?

Lack of confidence and anxiety are closely related and this seems to have particular implications for pupils' mathematical performance. In recent years considerable attention has been paid to this relationship. As Elizabeth Fennema (1980) succinctly puts it: 'One tends to do those things that one feels confident to do and to avoid activities that arouse anxiety.' For people who have always been good at mathematics it is sometimes difficult to understand the feelings of those who have always felt at best uneasy and inadequate and at worst totally numbed when confronted with mathematical problems or tests. Feelings of anxiety have been described in a number of ways: the mind 'going blank', 'sweatiness or coldness of hands', 'butterflies in the tummy', and even 'dizziness' and 'faintness'; and it is more likely that girls rather than boys will admit to having some of these feelings (Buxton 1981). You may well have had some of the anxiety feelings mentioned here although your maturity and experience may help you to suppress or deflect such feelings. Pupils, however, cannot rely on that experience.

Anxiety about mathematics seems to be a major reason for girls' under-achievement since they frequently 'withdraw' from the subject – in more ways than one – as a result of these feelings. Reassurance and sensitive handling by the teacher could help considerably here, but in order to provide such support it is necessary to be aware of the underlying causes of anxiety. Inevitably, there are numerous factors (social, cultural and psychological)

43

which may contribute to anxiety about mathematics. Some are beyond the influence of the individual classroom teacher (for example, the media representation of the 'swot'); nevertheless, there are strong indications that the anxiety that pupils feel towards mathematics is often exacerbated by the teaching style and classroom organisation of the maths teacher. Though there have been recent changes in pedagogical approach, traditionally maths teachers have relied on competitive activities, frequent assessment and emphasis on speed of work as a basis of their professional practice. Unfortunately, these particular elements are most likely to increase anxiety in pupils, and they therefore need to be handled sensitively. These issues are discussed in greater detail in chapter 3.

What can be done?

If teachers are aware of the pressures that girls and some boys experience particularly during adolescence, this should enable them to provide sensitive support, so helping to improve attitudes towards mathematics.

(1) Chapter 1 provides some indication of why mathematics is such a 'male subject'. Putting into practice some of the strategies suggested there could help girls to become more interested and *could* improve all pupils' perceptions of successful mathematicians. This in turn should promote more positive attitudes in girls and hence perhaps better performance in the subject.

(2) Providing suitable role-models could also help to improve pupils' images of successful mathematicians – possibly leading to increased motivation and success. For example, good women maths teachers often find themselves teaching younger children and lower abilities. If you fall into this category, offer to teach older children and examination classes in order to promote the image that girls can also be successful mathematicians. Provision of sensitive support (such as this might offer) has been shown to be a crucial factor in girls' continuing study of mathematics. You could also invite female visitors from industry, higher education, and so on, to provide evidence that attractive and popular girls can and do succeed at mathematics.

(3) The way that teachers organise their classes and the type of teaching and learning styles employed have considerable effect on pupils' confidence and anxiety. For example, all pupils' self-esteem and confidence can be raised by frequent praise and by ensuring a degree of success. Situations leading to student anxiety can be avoided if the teacher talks in a calm or reassuring voice, or helps anxious pupils privately. Be aware that students demonstrate anxiety by time-wasting, for example by pleading tiredness, by not having correct writing materials or by disrupting the work of the class. Be firm in dealing with this sort of behaviour. (See also chapter 3.)

(4) Discuss your pupils' attitudes with them whenever appropriate and challenge any stereotyped views when they are expressed. Barbara Binns in *Girls into maths can go* describes how attitudes in her third-year class improved perceptibly as a result of such discussions.

2.2 Teachers' attitudes and expectations

> Some teachers, you'll ask them in (say) technology or maths, you'll ask them something and ask how to do it but they would not just tell you how to do it, they'll do it for you and I dislike that.   (comment from a fifteen-year-old girl)

When speaking with pupils it is evident that teachers are of enormous importance to their schooling experiences, both in the everyday organisation of classroom life (this will be dealt with in detail in chapter 3) and also in teachers' perception of what their pupils can and cannot do. For example, have another look at the responses to the questionnaire. Did those pupils who agreed with statement 18 (I think my maths teacher thinks I'm stupid) have more negative attitudes in general than those who disagreed with the statement?

In this section we take a brief look at the expectations and perceptions of teachers and consider what influence these have on their teaching and thereby on their pupils' performance. Do you think that teachers have different expectations of girls and boys? The following activity invites you to explore your views.

Activity 2.9

This activity is based on one carried out by Judith Whyte (1985) as part of the Girls into Science and Technology project (GIST). It requires several groups of three or four people; you could do it with colleagues or with pupils.

(i) Prepare several copies of figure 2.4. Cut the copies so that the two case studies are separated. Give each group only *one* of these case studies. No group should be aware of what any other group is doing and each case study should be considered by at least one of the groups. Ask each group to reach an *agreed consensus* on

   (a) what the student is likely to be doing one year after leaving school (so that the first decision here is at what age the student would be expected to leave school);

   (b) what the student is likely to be doing when 30 years of age.

(ii) Compare what the groups have to say about Denis and Denise.

Comment

When this activity was tried by groups of teachers they felt that it successfully demonstrated the differential stereotyped attitudes held towards girls and boys. For example, Denise was seen by some teachers as a potentially successful and interesting person whereas Denis was condemned to a tedious working life with conventional responsibilities. (These judgements were made as a result of the O-levels on the profiles.) This group, considering Denise, felt that the combination of the four subjects indicated ability and potential, whereas the same results for Denis were seen as unexceptional and indicative of a rather unacademic plodding character! On the other hand, another group of teachers decided that Denis was likely to be more successful and to have a more interesting life than Denise because he would probably qualify for an apprenticeship whereas the girl was more likely to go straight into office work. Whatever their views on these pupils' likely experiences immediately on leaving school, most teachers felt that Denis was more likely than Denise to realise his potential. It was felt that he was more likely to go into further education, and he was seen as having considerable responsibilities by the age of 30.

Figure 2.4 Two case studies (adapted from Whyte 1985)

Case study 1

Denise Johnson, aged sixteen years, has the following characteristics:

1. well-liked by staff and pupils;
2. good looking;
3. has four O-levels, all with good grades, in
     English
     Maths
     Chemistry
     Geography
4. interested in 'helping people';
5. likes school, but
6. would like to start earning money as soon as possible.

Case study 2

Denis Johnson, aged sixteen years, has the following characteristics:

1. well-liked by staff and pupils;
2. good looking;
3. has four O-levels, all with good grades, in
     English
     Maths
     Chemistry
     Geography
4. interested in 'helping people';
5. likes school, but
6. would like to start earning money as soon as possible.

It was felt that Denise would probably be working part-time by the age of 30, fitting in routine office work around her home life.

This exercise is interesting because it shows that teachers often tend to expect far less of girls than they do of boys. Do you agree? Many teachers initially deny that they have such different expectations and are surprised by what they discover here.

(*Note.* You might also like to try this activity with (say) Asian names in place of Denis/Denise. This could lead to an interesting discussion about teachers' perceptions and expectations of pupils from different cultural backgrounds. Many teachers have higher expectations of some groups of girls – and lower expectations of some groups of boys – because of their cultural background.)

So teachers do seem to have different expectations about the behaviour and performance of girls and boys. They tend to have lower expectations of success for girls and make less academic demands of them compared to boys. Girls tend to be valued for neatness, conformity and good behaviour while boys are commended for exuberance, excellence and creativity. Rosie Walden and Valerie Walkerdine (1985) claim that this stereotyping of girls as neat, hardworking and good at computation, but not as imaginative and 'mathematical' as boys, is a self-fulfilling prophecy.

The GIST project found that teachers in general did not *believe* that they ever treated girls and boys differently. Gender differences in pupil performance were seen as usually linked to extra-school factors such as social expectation or the 'natural' differences between the sexes or youth culture (Payne, Hustler and Cuff 1984), but research has shown that teachers *unwittingly* confirm stereotyped views – often by being firm proponents of traditional values. For example, in the words of a primary head:

> My secretary is a woman and thank God for that. My teacher in charge of football is a man . . . in charge of netball is a woman. My cook is also female and the joiner who fixes our repairs is a man. How traditional I hear you cry! So what is wrong with being traditional? (Schools Council 1983)

At pre-school and primary levels, many teachers value male and female pupils equally but describe their typical behaviour very differently. In a study of primary teachers, Phil Clift (1978) found that the teachers 'saw the girls as being more sensible, obedient, hardworking, cooperative, quick, mature, bright, likeable . . . whereas boys were deemed more excitable, talkative, needed more supervision and attention' (see also Hazel Taylor's article (1986) in *Girls into maths can go*). Primary teachers appear to prefer teaching girls – perhaps just because they tend to conform to teachers' demands and are easier to relate to and control (for example, see Clift 1978). At secondary level, however, evidence suggests that there is an about-turn and teachers prefer to teach boys rather than girls in the belief that boys are likely to be more interesting and critical and also – of immense importance – that boys' education is more important than that of girls (Ricks and Pyke 1973). So it appears that teachers' views of their pupils' achievement are similar to those of the pupils themselves. Do you agree?

Activity 2.10

What sort of behaviour do you tend to reinforce in girls and boys? Take the mathematics reports of one or two classes. Remove any references to the pupils' names and blank out all pronouns.

(i) Ask a couple of your colleagues to identify whether each report refers to a girl or to a boy.

(ii) Ask them to note down what specifically characterises girls' or boys' reports.

Comment

When this activity was carried out at a mathematics workshop, most people correctly identified those reports referring to girls and those to boys. The criteria which were used for identification were:

- language: words like 'charming', 'neat', 'gossip', 'chatter', 'diligent' were applied to girls whereas words such as 'disruptive', 'demanding' were applied to boys;
- comments about presentation of work and diagrams were largely applied to girls;
- comments about non-completion of homework were frequently made of boys.

Were the results of your report activity similar? Did your colleagues identify similar criteria? Do you notice any other patterns here; for example, do you tend to reinforce different types of behaviour in pupils from different cultural and class backgrounds?

What implications do these studies on teachers' expectations have for girls' performance in mathematics? First, it seems that until comparatively recently most teachers have held views as stereotyped as those of their pupils, parents and society at large. Though the provision of equality of opportunity for their pupils lies at the heart of their professionalism, they seem to have been unaware of the possible impact of their own stereotyped views on the performance of their pupils. They have had little idea that, because of the unintended educational consequences of their everyday practice, they might well have actually discouraged or even prevented girls from moving into areas of study or employment traditionally orientated towards boys and men.

What can be done?

Because they are in a position to shape pupils' experiences, teachers have a crucial role to perform here.

(1) This section has shown that it is important for teachers to have high expectations for all pupils. Stereotyped views and expectations are often reflected in the way that teachers organise their classrooms and in the way they interact with their pupils. Reflecting on all aspects of classroom behaviour – and making appropriate changes to the teaching approach as a whole – can be of positive benefit to all pupils. The discussion in chapter 3 is particularly relevant here.

(2) More generally, taking an active interest in equality issues should help teachers to reach a better understanding of the influence that they exert over pupils, and this in turn could lead to changes in practice. Possible courses of action include:

    (a) reading the available literature on gender and race issues;

    (b) taking part in formal discussions about equality issues at staff and departmental meetings and also in informal discussions in the staffroom – sharing experiences, problems and strategies with colleagues encourages teachers to think seriously about *all* their responsibilities;

(c) discussing the issues with pupils and challenging stereotyped views;

(d) using in-service packs (such as this) or attending a suitable in-service course.

2.3 Parents' attitudes and expectations

> If I hear another mother say how poor she was at maths in school, I shall walk out.
> (comment from a teacher after a parents' meeting)

Parents have a very important part to play at every point in their children's schooling, and support or lack of support from them can make enormous differences to the feelings of pupils about school and about particular subjects.

Parents may hold quite traditional views about their children's upbringing. From early childhood, for instance, they provide their sons and daughters with different and often gender-specific toys: for example, dolls and domestic toys for girls, sports equipment and war games for boys. It has been claimed that in encouraging their daughters to be quiet, pretty and charming – in contrast to their sons whom they want to be active, vigorous and exploratory – parents unwittingly restrict the mathematical development of girls. For example, their spatial skills may not be as fully extended as those of their brothers.

Activity 2.11

(i) Which of the following games and activities, said to encourage spatial skills, are most likely to be played by girls and which by boys?

- building blocks, construction toys, and model building;
- electronic/computer games, chemistry sets, microscopes;
- strategy games such as chess, draughts and mathematical and jigsaw puzzles;
- ball games of all kinds, including billiards and snooker;
- physical and athletic activities;
- practical and craft activities – elementary carpentry, cooking, sewing and pattern making.

(ii) To what extent do you think children's mathematical performance is affected by the toys they have and the games they play at home?

Comment

Most of these games and activities tend to be played by more boys than girls, suggesting that it is not surprising that boys tend to develop better spatial skills than girls. Furthermore, the Cockcroft Report (1982) indicates that it is not only the games and toys with which children play but the way they are encouraged to behave which has implications for their mathematical future. Boys more than girls are encouraged to be independent, to experiment and to solve problems, all important characteristics of able mathematicians. To what extent do you think that the toys and games indicated above encourage the development of these skills?

It is not only in their encouragement of particular games and activities that parents can shape and reinforce stereotyped views; they can also have considerable influence when it comes to career and option choice at secondary

49

level. For instance, Alison Kelly *et al.* in *Girls into maths can go* found, when they asked parents to rate the suitability of various jobs for their children on a 1–5 scale, that parents had quite different occupational aspirations for girls and boys. They were happy to see their children in traditionally stereotyped jobs – i.e. their daughters as nurses, secretaries, social workers and hairdressers, and their sons as engineers, electricians and draughtsmen – thus reflecting, presumably, their own career aspirations and achievements. There is also some evidence (see Millman and Weiner 1985) that parents are more likely to *coerce* boys into useful career subjects while many girls are left to choose the subjects they like. So, for instance, many boys are encouraged to take physics or maths in addition to other science or technical subjects, though this rarely happens with girls.

Although we have been quite critical in our analysis of how parents' expectations influence their daughters' mathematical achievement, parents can (and often are) very supportive of their children's education and many would be horrified if they became aware of the restrictions they were unwittingly imposing on their daughters' futures. For example, Alison Kelly *et al.* found that most parents would not oppose their children getting a job stereotypically associated with the opposite sex; it is just that they do not envisage this happening.

There is another important parental factor – the parent as a role-model. The quote at the beginning of this section indicates that it is quite socially acceptable to admit dislike of mathematics and inability to do it. It does not seem to cause 'loss of face' – unlike, for instance, admitting difficulties with reading and writing. Research (see Russel 1983) has shown that girls are far more likely to study maths to a higher level if both their parents like it and are good at it. Moreover, fathers' high level of maths education significantly relates to their daughters' choice in taking maths A-level courses.

Activity 2.12

Look back at the responses to question 22 on the questionnaire. Whom do your pupils turn to for help at home? Can you detect any patterns?

Comment

If pupils say that they tend to ask their mother then this probably indicates that the mother is mathematically confident. Teachers in workshops have suggested that mathematically confident mothers are likely to have mathematically confident daughters and that mothers who lack confidence in mathematics are likely to affect adversely their daughters more than their sons. Do your results support these suggestions? You may also be able to identify cultural differences here. Patterns which apply to white pupils may not be relevant to black pupils.

What can be done?

Parents may well help to shape their children's mathematical experiences by the toys they buy, the behaviour and activities they encourage, their often stereotyped views about their children's occupational choices and as role-models. Nevertheless, many of them are also very concerned about equality issues. They therefore need to be presented with information about the importance of mathematics for all their children and advice on how to encourage them to study it seriously. There are several ways of doing this.

(1) The Equal Opportunities Commission (EOC) has produced a general guide and option choice in leaflet form entitled *Subject options at school: a positive choice at 13 has a positive effect for life* which is clear, succinct, accessible and comes free of charge from the EOC (see the additional information at the end of the chapter). Booklets produced by schools, whether on specific subjects or for general use, can be very useful in getting information across to parents.

(2) One of the best ways of raising parents' awareness is by organising evening sessions in which parents can discuss with teachers the issues relating to the education and future career prospects of their daughters. It is best to build the sessions around specific points of interest, to persuade parents on a cold evening or after a busy working day that it is worth coming out of their homes. In our experience, a session with 'career' and 'mathematics' in the title is likely to be far more attractive than one more generally directed towards equal opportunities. Care should be taken to allow sufficient time for questions and answers, and for parents to express their doubts and anxieties. Bearing this in mind, some of the topics in figure 2.5 might form the basis for useful discussion.

(3) An activity which demands somewhat more organisation and resources, but which has proved most valuable in improving parents' confidence in their own mathematical ability, has been the establishment of a series of

Figure 2.5 Points to raise at parents' evenings (adapted from Barnes, Plaister and Thomas 1984)

A Is there a problem?
- Sex differences in participation rates in mathematics and science
- Evidence of girls' lower achievement in some types of mathematics questions

B Why is it important?
- Mathematics as a 'critical filter' in the job market
- The increasing use of mathematics in all branches of society
- Job obsolescence and the need for flexibility in a rapidly changing world

C The changing roles of women and men in society
- The lifestyle of the average woman today and in the past
- Married women in the workforce
- Changes and variation in family structures
- The need for a job for security and independence – especially in times of high unemployment
- The importance of a job that is interesting and fulfilling
- Different cultural patterns and expectations

D What does research say about causes?
- The lack of evidence that biological differences between the sexes are the cause
- Sex-role stereotyping and its effect
- The importance of role-models
- Expectations – the self-fulfilling prophecy

E How can parents help?
- Avoid gender and cultural stereotyping
- Provide, where possible, positive role-models
- Discuss feelings about mathematics
- When buying toys, consider 'spin-off' effects
- Talk about career plans and expectations

sessions for parents on some of the topics being currently taught to their children in the school maths curriculum. About one such course, particularly for mothers, Helen Roberts and Alan Graham write:

> The aim of our *Sums for mums* project was to try a course of basic maths with a group of mothers. We chose mothers rather than [both] parents, because as Cockcroft points out, there is particular cause for concern about the achievement of women and girls in this area. But secondly, and just as importantly, it seemed to us that parents, and particularly mothers, are possibly the greatest under-used resource in education . . . We therefore decided to pilot a course which aimed both to increase the participants' confidence about basic maths and which would enable them to help their children in this area.   (Graham and Roberts 1982)

2.4   Employment and employers

Activity 2.13

How useful and relevant do your pupils think mathematics is to their future lives?

(i) Look back at the responses to statements 9 and 17 on the questionnaire. What proportions of boys and girls (a) agreed with, (b) disagreed with or (c) were neutral about these statements?

(ii) The responses to these statements provide some indication of pupils' perception of the usefulness of maths in their future lives. Do any distinct patterns emerge in your results?

Comment

Several studies have provided evidence that girls tend to believe that mathematics is personally less useful to them than to boys. For example, Sheila Russel (1983) showed that although girls, particularly those who were more academic, often rated mathematics as their 'most liked' subject, they tended to rate biology as more useful. Moreover, Russel found that girls tended not to specialise in mathematics because it was not seen as useful or directly relevant to any of the careers presented as suitable for them. Perhaps they believed biology to be more relevant to their lives as mothers and in predominantly female occupations such as nursing. The APU (1981) similarly found that girls perceived mathematics as less useful to them than did boys. If it is true – as the APU suggests – that pupils are more likely to find 'useful' subjects more interesting to study, then there seems to be some possibility of change if girls realise that mathematics can be useful to them.

Activity 2.14

This activity invites you to explore your pupils' views about careers. Prepare copies of figure 2.6 for the pupils in your tutorial group or in one of your maths classes.

(i) In column (2) ask pupils to put a tick alongside those jobs which they think are definitely maths-related, a cross alongside those which they feel are definitely *not* maths-related, and when undecided to leave a blank.

(ii) In column (3) ask pupils to write a letter B alongside those jobs in which they think more boys will be interested, a letter G alongside those jobs in which they think more girls will be interested, and to leave a blank alongside any which seem to be equally attractive (or unattractive) to both sexes.

Figure 2.6   Some ideas about careers

(1) Career	(2) Maths related (✓) Not maths related (✗)	(3) More boys (B) More girls (G)	(4) Most money (✱) Poorly paid (O)
Accountant			
Administrator			
Air hostess/steward			
Air pilot			
Assembly-line worker			
Banker			
Beautician			
Builder			
Clerk			
Computer operator			
Computer scientist			
Dentist			
Dietician			
Docker			
Draughts(wo)man			
Engineer			
Estate agent			
Factory worker			
Gardener			
Hairdresser			
Insurance agent			
Laboratory assistant			
Lawyer			
Lorry driver			
Model			
Nurse			
Plumber			
Police officer			
Radio announcer			
Radiographer			
Sales representative			
Scientist			
Secretary			
Shop assistant			
Teacher			
Technician			
Telephonist			
Town planner			
Typist			
Vet			
Waiter/Waitress			

Others in which you are interested

(iii) In column (4) ask pupils to draw a star (✳) alongside those jobs where they think a lot of money could be earned, a circle (o) alongside the jobs which they think are most poorly paid, and to leave blanks beside the others.

(iv) Discuss the results with your pupils. Are they at all surprised by what they have discovered? Can they suggest any strategies for change?

Comment

Did the results confirm that your pupils have very stereotyped views? Did they tend to think that more girls would opt for jobs in clerical work, sales and the service industries whereas more boys would choose professional and managerial occupations? Were there any differences in the views of pupils from different cultural or class backgrounds? And did pupils discover any relationship between traditional male careers, the amount of maths required in the job and the amount of money that might be earned? You might like to draw pupils' attention to the results of activity 2.13 here; also, just for information you might like to point out that mathematics is needed in all the occupational areas listed in figure 2.7.

Figure 2.7 Career areas where maths is needed (Smith and Mathew 1984)

Accountancy	Food science and	Pharmacy
Accounting	technology	Photography
Actuarial work	Forestry	Physics
Advertising (preferred)	Geology	Plastics and rubber
Air Force officer	Health service	technology
Air piloting	administration	Police work
Air traffic control	Horticulture	Post Office postal officer
Animal nursing auxiliary	Information science	Printing technology
work	Insurance (preferred)	Psychology
Architectural technician	Landscape architecture	Public relations
work	Leather technology	(preferred)
Architecture	Marketing (preferred)	Purchasing and stock
Astronomy	Market research	control
Audiology technician work	Mathematics	Quarrying
Baking technology	Medical laboratory	Radiography/
Banking	science work	radiotherapy
Biochemistry	Medical physics	(or physics)
Biology	technician work	Recreation
Brewing technology	Medical records work	administration
Building technician and	Medicine and surgery	Royal Air Force officer
technology work	Merchant Navy deck	(men and women)
Building societies work	officer	Royal Navy officer (men
Cardiology technician work	Merchant Navy	and women)
Cartography	engineering officer	Statistics
Ceramics technology	Merchant Navy radio	Surveying
Chemistry	officer	Surveying technician
Computer programming	Metallurgy	work
Dentistry	Meteorology	Teaching
Dietetics (preferred)	Midwifery	Textile technology
Dispensing optics (or	Neurophysiology	Town and country
physics)	technician work	planning
Dyeing	Operational research	Trading standards
Economics	Ophthalmic optics	administration
Engineering	Organisation and methods	Travel agency
Environmental health	and work study	Veterinary work
Estate agency and	Orthoptics	Women's Royal Army
auctioneering	Paint technology	Corps officer
Factory inspection	Patents	
Fashion design and	Personnel work	
production	(preferred)	

Pamela – interview with a bank manager:
'He said that day-release (to attend a college of further education) was mainly for men, for those men who wanted to become bank managers. He said he discourages women because they tend to leave and have babies and break their career!'

Diane – interview with engineering firm for apprenticeship – 20 minutes in length: '. . . he made it clear that he didn't think I would get the job and that he didn't want me to get it. He said "we have never had a girl here yet". The atmosphere was very tense.' [A boy from the same school – with lower educational qualifications – had a 45-minute interview and was offered an apprenticeship.] (Bennet and Carter 1981)

The pupils mentioned in these two short extracts were both applicants for jobs where mathematics and science qualifications were necessary pre-requisites.

It will probably come as no surprise to you that, for the most part, employers hold similar assumptions about the roles of women and men as do most parents, teachers and pupils; and many employers also hold views which disadvantage people on grounds of race. Traditionally, physical strength has provided much of the justification for the differentiation between women and men in job opportunities. Women tended to be excluded from most occupations and crafts in the nineteenth century and before on the basis of the 'natural' differences between the sexes and their lack of strength compared with men. The physical strength needed by women to rear families or to 'take in laundry', for instance, was largely discounted. However, times are changing and physical strength is now less often used at craft and manual levels where mechanical handling equipment has reduced the need for muscle power. Moreover, employers are under some pressure, through the Equal Pay (1970), Sex Discrimination (1975) and Race Relations (1975) Acts, to provide equal opportunities and conditions for males and females and not to discriminate on racial grounds. For the more academic, the professions are now ostensibly open to women though it is still very much more difficult for them both to enter and to reach the higher levels. In fact in a survey conducted in 1978 only 2% of all Professors, 9% of Medical Consultants, 1% of Civil Engineers and less than 1% of Bank Managers were women (see Miller 1981).

It is now a fact that most women work outside the home, whether single or married. In Europe, women are likely to be out of paid employment for a total of only seven years – on average – while they are rearing children (see Joshi, Layard and Owen 1982). However, women are effectively penalised as a result of child-caring breaks by being restricted in career choice to a narrow range of occupations and are invariably more poorly paid than their male equivalents.

There is plenty of evidence about the stereotyped views of employers. For instance, as part of the Schools Council Sex Differentiation Project, a Leicester careers teacher attempted to raise awareness of issues of gender amongst employers by asking them to provide information for a survey he was carrying out on equal opportunities (see Millman and Weiner 1985). His employers' survey revealed that the employers in his area held views which were in direct conflict with the school's equal opportunities policy. For example, he draws attention to 'the factory manager who claimed "women are the best machine operators for two reasons: they can work more quickly on the machines and they only want money to support their husbands' incomes"', and many people can recall similar anecdotes.

Not unexpectedly, pupils tend to have similarly stereotyped views about careers. For example, as part of the Leicester survey mentioned above, the teacher administered a questionnaire to fourth- and fifth-year pupils on the relationship between subject choice and career aspiration. While his analysis did not reveal a significant difference between career aspirations among the more able girls and boys, he found that the boys appeared far more likely to go for a career in Industry while the girls tended to choose the Arts and Social-Welfare/Public employment (although this may well have reflected job opportunities in the area at the time). At the age of sixteen he found that boys were likely to enter careers on a structured training programme (60% of the boys gained apprenticeships compared with only 4% of the girls) while girls entered what he loosely defined as 'office work' jobs, with very little equivalent training. He also found that, given equal qualifications, boys had a higher chance of achieving their job aspirations than girls and able girls frequently found their career course diverted by their employers' attitudes. This research therefore showed that it was not sufficient to raise girls' aspirations unless employers' attitudes were also challenged and changed.

Stereotyped views about careers must have implications for mathematical performance. If pupils take up subjects which they see as useful in adulthood and if they see employers appointing only boys to jobs using maths and science, then girls are unlikely to look at mathematics and science as appropriate subjects for them to take. In their view, what is the point in studying maths to a high level if it will be of no real use to them? With increasing unemployment many boys may feel the same – particularly if they are disadvantaged because of their experience of negative attitudes towards their race or class.

What can be done?

The views of both pupils and prospective employers need to be tackled here – although the former is rather easier than the latter. As mentioned earlier, teachers have considerable influence on pupils' attitudes, and so can encourage them to broaden their views about suitable careers for girls and boys from all backgrounds. In so doing, girls could well come to see maths as a more useful subject to them and their performance might consequently improve.

(1) One way of encouraging pupils to broaden their views might be to provide careers advice in the early years of secondary schooling – long before the advent of option choice – so that pupils understand the implications for their chosen career (or course) of not doing well in mathematics.
(2) Also the case should be made that mathematics can be needed in traditional 'female' jobs such as hairdressing, nursing and sales work, but that this often goes unrecognised.
(3) Although it is difficult to talk in terms of career choice at a time of high youth unemployment and few occupational openings, the suggestions in figure 2.8 are still worth making to all pupils.

Turning now to employers' attitudes: at a time of high youth unemployment teachers have found it quite difficult to persuade employers to establish a 'fair' policy on recruitment and promotion since the employers 'call

Figure 2.8   Choosing a career

> ■ Do not choose a job or career just because it is thought appropriate for girls or suitably 'feminine'. These kinds of jobs generally draw low pay and have low status.
> ■ Assess your capabilities and interests realistically and do not *underrate* yourself.
> ■ Do not select a job that looks easiest or pays most in the short term. It may become the most boring and low paid job in the long run.
> ■ Look for advice about jobs from a wide range of people – do not rely on one person alone.
> ■ When looking for a job, take into account:
>   (a) whether it will be interesting and/or worthwhile;
>   (b) whether there are opportunities for training and promotion;
>   (c) whether it offers a good future.

the tune'. Some pressure is currently being applied by the Manpower Services Commission (MSC), which has close connections with industry, finances much of the current on-the-job training and has an official commitment to the promotion of equality of opportunity. The MSC is in a position to generate greater awareness of equality issues amongst employers but what can teachers do?

(4) The Leicester survey (see Millman and Weiner 1985) suggests one possibility. Here, the employers were asked to provide information to the school in a number of areas – not only to help the school with its careers work but also to use the respect and trust established between local companies and the school to negotiate changes in the attitudes of the employers. Questions were asked of the employers about the following:

■ the number of female/male employees;
■ the types of jobs and areas of responsibility undertaken by female/male employees;
■ the reasons for any existing sexual division of labour;
■ the company's policy on equal opportunities.

The implications of the results were then discussed with the employers. Admittedly, there were mixed responses to this survey and to a subsequent invitation to the employers to attend school events. Nevertheless, it was felt that at least awareness of the issues had been raised among some of the local companies. You might therefore like to undertake a similar survey. Your survey could incorporate issues of race as well as the gender issues discussed here.

2.5   Summary and reflection

In this chapter we have described the variety of ways in which girls' performance in mathematics is shaped by the attitudes of their peers, teachers, parents and employers as well as by girls themselves. We have offered a number of suggestions through which teachers could begin to tackle the massive task of changing attitudes.

Activity 2.15

Jot down in the table below the aspects of the problems discussed in this chapter which are most relevant to your pupils or to your school. Beside each of these, note the strategies that you feel you can most effectively introduce to improve the situation either now or in the future.

Aspect of the problem	Strategies for tackling it

Further reading and additional information

- Articles by L. Joffe and D. Foxman, H. Taylor, A. Kelly *et al.*, Y. Benett and D. Carter, in *Girls into maths can go*
- The EOC booklet *Subject options at school: a positive choice at 13 has a positive effect for life* is available from
  The Equal Opportunities Commission,
  Overseas House,
  Quay Street,
  Manchester M3 3HN

3　In the classroom

A group of nine- and ten-year-old children were gathered on the carpet with their teacher. Together they were working at a mathematics problem on a small blackboard leaning against the wall. The teacher was called away for a few minutes. Within seconds, all the boys in the group had gathered around the blackboard, elbowing the girls away and blocking their view. When the teacher returned, she found the boys busily engaged in the task, the girls 'gossiping' on the fringes.

(observed in an ILEA primary school)

It was the start of the new school year in a secondary school. The boys and girls were lined up outside their first mathematics class. As the teacher supervised them filing in he said: 'Girls, sit at the back because mathematics is not such an important subject for you as it is for the boys.'　　　　(reported by a twelve-year-old girl)

Some children were told the story of the boat problem: two adults and two children want to cross a river but their boat will only hold one adult or two children. How do they get across? A boy solved the problem first. Eventually the teacher suggested he help the other children understand it. He first sat with his friend (another boy) who was using a box 'boat' and Cuisenaire people. The boy who had solved the problem was obviously dying to show his friend what to do. But he remained quite patient, allowing the other boy to handle the equipment and to think through the problem himself, only intervening when it was obvious that he was forgetting something crucial, such as 'Who will send the boat back?' However, when they were both asked to help one of the girls, they did not allow her anything like the same amount of thinking space. They hardly allowed her to touch the boat and people. They manipulated the box and blocks and discussed the moves. The girl made some vain attempts to push them off. When the boys finally smiled and stated 'problem solved', she just sat and stared, having taken no active part.

(ILEA 1984)

All of the above incidents have taken place since 1983. You might feel that they are isolated and unusual, but discussions with teachers and pupils indicate that pupils frequently report such experiences. As we have already pointed out, the reasons for girls' under-achievement in mathematics and their withdrawal from the subject are extremely complex. Mathematics tends to be seen as a 'male domain' in more ways than one and this partly accounts for girls' negative attitudes towards the subject and hence perhaps for their lower mathematical performance. But it does not totally explain the types of feelings that pupils often associate with mathematics – for example, feelings of panic, anxiety and pressure. As mentioned in chapter 2, such feelings could well result from the way the classroom is organised and from the teaching and learning styles that are employed.

Activity 3.1

Look back at the opening anecdotes. Jot down any aspects of classroom organisation and interaction that you feel might be relevant to the behaviour of the pupils in these situations.

Comment

The first anecdote describes a situation which frequently arises as a result of pupils' own perceptions (that mathematics is a male domain). The danger lies in the possibility that the teacher, on returning to the group, unwittingly makes assumptions about interest and motivation and, in an attempt to discourage apparently irrelevant behaviour on the part of the girls, chides them for failing to participate. On the other hand, an awareness of the ways in which boys can enthusiastically 'shanghai' a discussion, might lead the teacher to chide the boys for excluding the girls from the action and thus settle the group back to working together as they had been doing.

The second anecdote is an example of the kind of gratuitous remark which forms the basis of much social exchange with pupils. Here the teacher is organising his class in a way which reinforces any stereotyped perceptions that the pupils themselves might have.

The third anecdote provides an example of the assertive behaviour demonstrated by many boys particularly in the presence of girls. When pupils are observed it is not unusual to find boys ridiculing girls, abusing them verbally, or harassing them physically. Too often, such perceptions of the boys remain unchallenged by the teacher even when they are reprimanded for the unkindness of the behaviour. The Cockcroft Report (1982) says that 'in such circumstances girls are likely to receive the message that they are not expected to perform as well as the boys and to react accordingly.' The way that teachers interact with pupils – and indeed their whole philosophy about teaching and learning styles – can reinforce these expectations.

This activity draws attention to the fact that the learning of mathematics takes place in a social environment which itself influences that learning. Pupils not only learn the subject but they also learn about learning and pick up expectations relating to themselves as individuals and as members of a group. What pupils learn at school is often very different from what it is intended they learn. This chapter is about these 'meta-learnings' and the influence they have on pupils. Some that come to mind are aspects of school organisation and school life that cause girls to perceive their role as restrained in some way, so that they lack confidence in dealing with certain subjects; these are discussed in the first section of this chapter. Others may be grouped together under the general title of 'teaching and learning strategies'. These are explored in the next two sections under the headings 'Interaction in the classroom' (section 3.2) and 'Pupils' preferences for different teaching and learning styles' (section 3.3). In each section, research findings about gender differences in behaviour will be examined and related to your own situation. You do not need to have read the research literature to be aware of differences in behaviour between girls and boys (and between girls and boys from different cultural and class backgrounds). What is important in terms of the learning of mathematics is how much these differences are initiated or exaggerated during classroom interactions and whether they can be influenced by teacher actions.

Whether you teach a mixed-sex, or a single-sex, class you will be aware of differences in how pupils behave. However, those interested in gender issues in learning have noted that different criteria are used to judge the 'appropriateness' of the same behaviour in girls and boys. For example, a competitive, sportive, physically active girl might be called a 'tom-boy' thus identifying her as behaving differently from the norm expected for girls. Equally, a quiet, studious boy interested in books might be labelled 'sissy'. Such labels are, of course, unhelpful because of the way they preclude an appreciation of the person's individuality and obstruct development of natural interests. But, curiously, it is not only in mixed-sex classes that this kind of behavioural stereotyping has been noted. Those who teach in single-sex classes have observed that there are boys who fill the 'feminine' roles – being quiet, conforming to teachers' expectations, being neat and tidy and hardworking – or girls who fill the 'masculine' roles – being confident, assertive, competitive and often unruly. Of course, in the real classroom nothing is as clear-cut as this. However, it has been observed that on a number of continua for different behaviours, the 'feminine' version lies at one end, and the 'masculine' at the other. For example, in considering pupils' responses to teacher questioning, the continuum might go:

←――――――――――――――――――――――――――――――――→

demonstrating confidence   lacking confidence

In general, more boys seem to lie to the left, more girls to the right. In a single-sex class the pupils are also likely to span the continuum and thus there will be some confident girls in a girls' class and some insecure boys in a boys' class. Do these pupils get treated differently by the others? Are your expectations of them different? Do they get 'labelled' differently because of their behaviour rather than their sex? If yours is a single-sex class you might like to watch for pupils filling different roles, which reflect sex-stereotyped norms, when trying the activities in this chapter.

It may also be the case that girls (and boys) from different cultural and class backgrounds can be found clustered in different places on the continuum. This too can be borne in mind when working through the chapter.

Activity 3.2

First of all, in order to place the chapter in context, it is important to gather some information about your pupils and to reflect upon your patterns of interaction with them. (This activity is adapted from one in Barnes, Plaister and Thomas (1984).)

(i) In the first column of figure 3.1 record the names of all the pupils in one of the classes that you teach. Record the sex of each pupil in column (2).

(ii) (a) In column (3) place a tick against each of the five pupils that you feel you know best, and a cross against each of the five you feel you know least.

(b) In column (4) use ticks to identify the five most confident pupils and use crosses to identify the five least confident pupils in the class.

(c) In column (5) use the same method to identify the five most able pupils and the five least able pupils in the class.

(d) In column (6) use the same method to identify the five most responsive pupils and the five least responsive pupils.

(e) In column (7) use the same method to identify the five pupils with the highest level of attention-seeking and the five pupils with the lowest level of attention-seeking.

(f) In column (8) you might like to categorise similarly some other aspect of behaviour.

(iii) Do any patterns emerge in your results? Are there any surprises for you?

Comment

Did you notice anything about the order in which you wrote down the pupils' names? Were the assertive and/or confident learners the ones you remembered first? Are they the ones you feel you know best? Which pupils do you know least about?

Figure 3.1 Profile of class

(1) Pupil's name	(2) Sex	(3) Knowledge	(4) Confidence	(5) Ability	(6) Level of respon- siveness	(7) Level of attention- seeking	(8) Other

Many teachers find that the pupils they know best are those that are either the successful or the unsuccessful learners; the ones they know least are those who are described as 'average' or 'quiet'. In mixed-sex classes such pupils are frequently girls; as one teacher remarked, 'I know least the average, undemanding girl.' How do your findings compare with this? Do you notice any patterns connected with gender, race or class?

Do you think that your expectations of your pupils are independent of their performance or are the pupils you expect to succeed the ones who do? In a study in the USA (see Rosenthal and Jacobson 1968) researchers deliberately misinformed a teacher about the abilities the children in the class had displayed in some tests. When the researchers returned some months later the children's performance more closely matched the misinformation than the reality; for example, an average child labelled 'bright' was performing at a higher level than previously. Does this suggest a self-fulfilling prophecy?

3.1   The unintended effects of school organisation

A school is a social organisation and its structures will reflect the conditions and norms that are current in the society at large. Thus it is not surprising to find that the sex and race distribution amongst senior management in schools does not reflect the distribution across all grades. For example, in ILEA secondary schools, 51% of full-time teachers are women of whom approximately 60% are on scales 1 and 2 – whereas 47% of the male teachers are on these lower scales. In primary schools, 80% of the teachers are women of whom about 77% are on scales 1 and 2 – compared with 49% of men. And the distribution in ILEA is better than the national distribution (Burton and Townsend (1986) in *Girls into maths can go*).

Inevitably, the sex and race distribution amongst teachers can influence pupils' attitudes as part of the hidden curriculum (i.e. the learning that takes place in school but which is not part of the formal curriculum). Pupils cannot avoid being taught by the (often white male) senior teachers or heads who sometimes quite overtly reinforce the equation of mathematics' future importance and masculinity – as indicated in the second anecdote at the start of this chapter.

There are many aspects of the hidden curriculum. For example, girls' schools are usually under-resourced for the teaching of laboratory-based subjects compared with boys' schools. Even where the resourcing is roughly the same – as in mixed-sex schools – timetabling can impose other inequalities, such as when a subject seen as 'male' (e.g. physics) is timetabled against a subject seen as 'female' (e.g. home economics). Since the evidence suggests that studying mathematics is helped by studying other maths-related subjects, there can be a discriminatory effect on girls' performance from such a timetabling distribution. The way the classroom is organised also influences pupils' attitudes. For example, girls and boys are frequently divided by the register and by patterns of seating and this can affect their expectations – as exemplified by the following anecdote:

> David, aged five, came home during his first week in school and announced, 'Boys are best.' When asked to explain, he said: 'Well, they must be because they are first on the register.'

Activity 3.3

Jot down two or three other aspects of the hidden curriculum that may influence girls' performance in mathematics.

Comment

We mention a few here – you may well have thought of others. As already discussed in chapter 2, the attitudes and expectations of teachers must have some influence on pupils' performance. Also, the way girls and women are portrayed in textbooks (see chapter 4) and the male domination of senior school management might convince them that high-status jobs are not within the scope of women. And boy-dominated school computer or chess clubs can convey to girls that certain activities are principally organised for boys and men. One claim is that the uniform girls wear may have an important influence on their eventual mathematical attainment. Girls usually wear skirts during some or all of their school life (either because of regulations, parental taste or juvenile fashion). By wearing skirts or dresses, it is suggested, propriety discourages girls from moving about freely and exploring space (for example, by climbing trees) in the same way that boys do. The unforeseen consequence of this could be that the development of girls' spatial awareness and skills may be restricted. This in turn could put them at a disadvantage when taking mathematics.

You may also be able to identify aspects of the hidden curriculum that could particularly influence the performance of girls and boys from different cultural or class backgrounds, and you might like to discuss these with colleagues.

What can be done?

Some of the aspects of the hidden curriculum mentioned in the discussion above are dealt with in other chapters of this pack (for example, attitudes are discussed in chapter 2, and bias in teaching materials is discussed in chapter 4). We concentrate here only on issues concerning school and classroom organisation.

(1) Teachers attempting to implement an equal opportunities policy have noticed that it is difficult to do so in the absence of a whole school approach. Alan Eales (1986) in *Girls into maths can go* draws attention to a number of organisational features which might be considered by the school as a whole. He advocates for example:

- wide discussion to establish a common frame of reference amongst staff and pupils;
- having the support of senior management;
- developing an integrated school policy;
- making the concerns public;
- actively involving the pupils in discussing the issues;
- attending to small organisational changes;
- seeing anti-sexism as good practice which enhances *all* pupils' learning.

(2) In particular, it might be a good idea to consider making small changes in the way the classroom is organised; for example, by reordering the register so that it is completely alphabetical (rather than first boys then girls), or reassessing the seating arrangements. Avoid reinforcing stereotyped behaviour in the classroom (such as girls always being asked to carry

messages) by, for example, deliberately asking both girls and boys to run errands.

(3) A more radical suggestion is to consider the possibility of segregating girls from boys – at least for the teaching of mathematics. Segregation as a response to sex-differentiated behaviour in the classroom was first reported upon by Stuart Smith (1983). He outlined the experiences at Stamford High School where an amalgamation of an academic boys' school with a less academic girls' school resulted in the observation that the girls were passive in class and under-achieving across the ability range. First-year girls were equal to boys when tested on entry but by the end of their first year had already fallen behind in test results as well as showing evidence of negative attitudes. By the fourth year, boys outnumbered girls by four or five to one in the two top mathematics sets. The few girls in these sets reported feeling uncomfortable in the masculine environment. They feared ridicule and were observed to be adopting a passive role in class. A two-year experiment was undertaken in which half the top-band students were assigned to single-sex and half to mixed-sex maths sets in their first year. When tested in November of the second year, the single-sex girls' performance was about equal to the mixed-sex and single-sex boys' and clearly superior to the mixed-sex girls', and this gap continued to increase. The girls' results are shown in figure 3.2.

Figure 3.2 Stamford results

	Mean scores on tests		
	October 1978 (initial selection test)	November 1979	February 1980
Single-sex girls' set	58.9%	55.1%	54.7%
Girls in equivalent mixed-sex set	58.1%	50.0%	43.9%

The October 1978 scores indicate that at the time of the initial set selection there was little to choose between the girls in either set. By February 1980, the average score of the girls in the mixed-sex set had fallen well behind that of the boys in the same set. In other words, these girls were conforming to the typical pattern for the school. The girls in the single-sex set, however, achieved a far better average score than the girls in the mixed-sex set and were only slightly below the average score achieved by the boys. Whereas nine of the sixteen girls in the mixed-sex set failed to achieve 40% in the February test, only four out of thirty-one girls in the single-sex set failed to obtain this score. Because of the success of this experiment the school changed to single-sex maths sets throughout the school, and as a result the number of girls taking and passing O-level maths increased, as did the number of girls who opted to take A-level maths.

Alan Eales (1986) in *Girls into maths can go* reports on a similar experiment conducted with fourth-year pupils. An interesting feature of this experiment was that it was done in the context of a whole-school approach to equal opportunities and the particular outcomes of the single-sex division could in part relate to the policy of anti-sexism throughout the school.

3.2 Interaction in the classroom

Researchers have suggested that active assertiveness and confidence, when adopted by children, are the characteristics necessary for full participation in the learning process (for example, see Eynard and Walkerdine 1981). Unfortunately, assertiveness and confidence can be interpreted, in the classroom, as challenges to the teacher's authority and can then directly affect the way in which that teacher perceives the task of motivating and controlling the class. Do you think that teachers respond differently to assertiveness and confidence in girls and boys? (You might like to look back at section 2.2.)

Perhaps one of the most obvious ways that teachers motivate their pupils lies in their use of praise and criticism. For example, if you feel that a pupil is a successful learner do you expect more and criticise more?

Activity 3.4

The next time you are marking pupils' exercise books, take a look at the types of comments that you tend to make. Jot down the types of comments you make when praising and criticising:

(i) successful learners,
(ii) unsuccessful learners,
(iii) 'average' learners.

Are there any implications in your comments, overall or in particular for the girls in your class?

Comment

Did you find that your use of praise and criticism was related to your expectations of pupils' performance? Were your comments influenced by the sex or race of the pupil? The literature suggests that boys receive more explanations, directions and extended assistance, and are encouraged to be more independent than girls. When criticised, boys seem to be singled out for bad behaviour, boisterousness or untidiness. Girls tend to be praised for behaviour such as neatness, tidiness and being conscientious. It has been suggested that consistent praise for neatly presented work leads girls to allocate high priority to good presentation; but this can rebound against them, as it is then easier to see and identify errors and their work usually carries critical comments from teachers which relate to the mathematical content. On the other hand, boys' work is frequently untidy and difficult to read and thus carries critical comments relating to presentation.

Such differential use of praise and criticism has been conjectured as one possible explanation for why girls relate criticism of their performance in mathematics to their lack of ability (something which is internal and which they see as unchangeable) whereas boys relate criticism to factors such as 'not making enough of an effort' or 'bad luck' (which could change). What is being said here is that it is not so much the *fact* of praise or criticism as its *substance* which will affect the way in which pupils interpret it.

In a large study in the United States, Carol Dweck and Ellen Bush (1976) investigated whether the content of praise/criticism related to intellectual qualities like good ideas/wrong answer or non-intellectual aspects like neatness/bad setting out and how this differed for girls and boys. They found that the positive and negative feedback on the intellectual quality of the pupils' work was roughly equally distributed between girls and boys. However, nearly two-thirds of all negative feedback to boys referred to non-intellectual aspects of their work. If only the negative feedback which was *work-related* was examined, roughly one-half did not reflect upon the boys' intellect. By contrast, the girls received almost no criticism for non-intellectual aspects so that nearly 90% of the negative remarks directed towards them referred to their intellectual quality. The teachers, it was noted, did not make frequent explicit references to the reasons for pupils' errors but, where they did, they were eight times more likely to attribute a boy's failure to insufficient effort than a girl's.

Do you think that your use of praise and criticism plays an important part in the way that your pupils view success and failure?

Activity 3.5

(i) Look back at the responses to statements 11, 15, 20 and 21 on the questionnaire provided in activity 2.2. What percentages of girls and boys responded in each way to these statements?

(ii) These statements provide some indication of how pupils perceive success and failure. Do any distinct patterns emerge in your results: overall, or with particular reference to girls and boys, or with reference to pupils from different age-groups or from different cultural and class backgrounds? Are there any surprises for you? You might like to discuss your findings with your pupils.

Comment

Do your results confirm that boys tend to attribute success to ability and failure to bad luck and lack of effort, whereas girls tend to attribute success to hard work and good luck, and failure to lack of ability? To what extent do you think that you reinforce these attitudes by the type of praise and criticism you use in comments and in classroom interaction?

There has been considerable discussion about whether the differences established between girls and boys in the way they explain their successes and failures have implications for their eventual mathematical performance. If, as has been suggested, boys do tend to explain failure in terms of changeable or controllable factors (such as luck and effort) in contrast to girls who see success in these terms but failure in the light of unchangeable or uncontrollable factors (such as lack of ability) then this has obvious implications for their motivation and attitudes to mathematics. It also has implications for their mathematical performance since it has been claimed (see again Dweck and Bush 1976) that pupils who attribute failure to uncontrollable factors show disrupted performance and decreased effort when confronted with failure or examination pressure – and therefore tend to avoid situations in which failure is likely. On the other hand, those who attribute failure to controllable factors show improvement in performance and increased motivation – and enthusiastically approach tasks that present challenges.

Turning now to more general aspects of interaction in the classroom, the Cockcroft report (1982) states that 'teachers interact more with boys than with girls, give more serious consideration to boys' ideas than to those of girls, and give boys more opportunity than girls to respond to higher cognitive level questions'. On the other hand, there is some evidence that this general observation does not apply to all pupils. For example, the Swann report (1985) comments on the negative nature of interactions experienced by some black pupils. What are your views?

Activity 3.6

This activity invites you to explore one or more aspects of your interaction with pupils in the classroom. You might like to try it out with the class you listed in activity 3.2. You will need the help of someone to act as classroom observer.

(i) Decide which of the following aspects of your interaction with pupils you would like to investigate:

    (a) where you spend most of your time;
    (b) the type of praise/criticism offered and to whom;
    (c) which pupils answer questions most frequently, and which pupils answer different types of questions;
    (d) which pupils talk most to the teacher.

(ii) You will need to arrange your classroom and decide where your pupils will sit in advance of the observation period. Indicate on the observation sheet in figure 3.3 the layout of the classroom (for example, the location of tables and so on). Allocate a square to each pupil according to where they will be sitting. Number each of these occupied squares and indicate whether the occupant is a girl or a boy with a 'G' or a 'B' in the top left-hand corner of the square. Try to indicate whether or not each pupil is confident in mathematics with a 'Y' or an 'N' in the top right-hand corner of the square. (If appropriate you may also like to make a note of each pupils' race.) Prepare one copy of the observation sheet for each of the selected observation tasks so that they can be completed and analysed separately.

(iii) Ask your observer to carry out the observation task(s) selected from the list in part (i), using the following notes for guidance.

    (a) *Where you spend most of your time:* record a 't' on each square which most nearly identifies the teacher's position in the room every 60 seconds.
    (b) *The type of praise and criticism offered and to whom:* record in the appropriate square 'p' for praise relating to intellectual aspects of pupil's work (e.g. 'you have a good idea there'); 'c' for criticism relating to intellectual aspects (e.g. 'that's a bad mistake'); 'np' for praise relating to non-intellectual aspects (e.g. 'well-presented example') and 'nc' for criticism relating to non-intellectual aspects (e.g. 'that work is too untidy').
    (c) *Which pupils answer questions most frequently and the types of questions answered:* record in the appropriate square whenever a pupil answers a question asked by the teacher. Discriminate between a factual answer by recording 'f' and an explanation by recording 'e'.
    (d) *Which pupils talk most to the teacher:* tally in the appropriate square *each time* a pupil speaks to a teacher. Each time a pupil *initiates* an exchange with the teacher, cross the tally (✗).

(iv) Discuss and reflect on your results. Compare your findings with your results to activity 3.2. Is there a relationship between how well you know a pupil and how much time you spend with her or him? Do you spend more time with assertive or confident pupils? And are there any implications for girls in your findings?

Notes

(a) During the development of this pack some teachers worked with a colleague and found it particularly informative where they paired up with someone of the opposite sex and observed in each other's classes. Other teachers used teaching-practice students as observers and some asked a member of the class to fill this role. Asking a member of the class led to some interesting discussions with the pupils about their own feelings and observations. If your school has a video-camera you might like to ask someone to video one of your lessons. You could then act as your own observer and carry out each of the observation tasks in turn. Also, if it is not possible to find someone to help you here, you could use a cassette recorder to tape your own lesson. Knowledge of your pupils' voices could perhaps help you to carry out observation tasks (i)(b), (i)(c) and (i)(d) yourself.

(b) It is not realistic to attempt more than one of the observation tasks at any one time, although if your observer will be with you for a long lesson period you may wish him or her to observe different aspects of your interactions with pupils (say) every ten or fifteen minutes.

(c) If pupils are not in their own seats at any time during the observation period, ask your observer to record the appropriate information in the square which most nearly describes their position and, if possible, to indicate the number of the 'home' square.

Comment

One teacher expected to find that she spent more time with boys than with girls, but this activity indicated that she spent an amount of time with each sex that was roughly proportional to the number in the class. Another teacher noted that he spent approximately the same time with first-year girls as boys. However, in the fourth-year class, he was spending less time with the girls and he commented that this result corresponded to a decrease in positive attitude which he had found in the attitude questionnaire (see activity 2.2) that he had administered to the same classes. What do your findings suggest to you?

In responding to the first observation task (i)(a) – recording the teacher's movements – an observer noted that the teacher had made the same number of visits to girls as boys *but tended to stay longer with the boys*. This is not surprising; frequently it has been established that the 'typical' boy – aggressive, competitive, often untidy, and frequently disdainful of 'girls' activities' – demands and is given the most attention. For example, in a small study of ten high-school geometry classes in the USA Joanne Becker (1981) found that if the teacher asked a question and a student called out the answer, the teacher would then direct her or his attention to that pupil. It is a very human reaction to relate to those who are showing signs of voluntary participation. It is the most assertive and responsive pupils who tend to participate in maths lessons and it should not be too surprising to find that it is they who receive the most assistance and encouragement. If, as research suggests, boys are more assertive than girls then it is likely that they also receive more attention than girls. Joanne Becker's findings seem to confirm this; she found that 67% of the instances where a pupil called out an answer and commanded the teacher's attention were with boys, and 33% were with girls. A small investigation carried out in two primary schools in Australia by Terry Evans (1982) also suggests that it is the assertive 'participators' to whom much of the teacher's attention is directed, and that these are most often boys.

Figure 3.3 Observation sheet

Length of lesson _____ Class _____ Teacher _____ Observer _____

Observation task _____

Conscientious pupils who work hard, are well-behaved and achieve well generally have *longer* interactions with the teacher – though less often than assertive pupils. But even with these pupils there are differences in the *type* of interaction with girls and boys. Boys tend to be praised for their work and are given assistance and explanations whereas girls receive more general encouragement.

So not only do assertive pupils appear to receive more of the teacher's attention, but it appears that there are different kinds of attention. For example, it has been suggested that teachers are prepared to wait longer for a response from a boy than from a girl, as illustrated by the following incident described by Elizabeth Fennema (1980):

> Mathematics class with the teacher moving around providing individual help:
> Teacher: Have you figured out the answer, Marcia?
> Marcia: Uh, no. Not yet.
> Teacher: Eric, how about you?
> Eric: I can't get it!
> Teacher: Come on, Eric. You can do it. What's the exponent?
> Eric: Oh, yeah, x to the fifth. I get it now.

The least assistance and the slowest teacher responses seem to be reserved for the quiet, neat, tidy, diligent, average girls (see Evans 1982). Do your results support these research findings?

Frequently, boys initiate interactions with the teacher by calling out and guessing. Joanne Becker (1981) found that sometimes girls do appear to try to redress the imbalance in interactions by initiating contacts with the teacher. However, even when teachers are aware of the inequality of time and attention that is given to boys compared with girls in mixed-sex classes, they find it extraordinarily difficult to achieve a balance (see Spender 1982a). Is this because they identify as normal, assertive behaviour in boys and passive behaviour in girls, and that their patterns of interaction reflect this?

Can you identify any other patterns in your results, for example with regard to race and class?

What can be done?

This section has drawn attention to some aspects of classroom interaction that could affect girls' mathematical performance. All of these are likely to be responsive to alert and sensitive handling by teachers.

(1) Once aware of the effects of praise and criticism and the distinction between internal factors like ability and external ones like effort, a teacher can monitor the use of interjections in class and the use of comments on written work to try to re-adjust the balance. One teacher commented that it was only as a result of being observed responding 'long distance' across the classroom to boys' questions that she became aware of how much the girls disliked asking questions and getting such a public response. She immediately changed her style and made sure that she always crossed the room and walked up to a pupil who wanted attention so that the exchange could be personal. The feedback that she had from the girls was very positive.

(2) Similarly, by becoming aware of the fact that quiet, average, undemanding pupils are sometimes repeatedly 'forgotten', teachers might consciously try to direct more attention to them.

(3) Being more conscious of the need to wait for responses and to encourage girls to experiment and try, could help teachers to explore mechanisms to make this possible. For example, observing unbalanced response rates in a class encourages the rethinking of the types of questions asked and the types of responses expected.

(4) Perhaps the most important realisation here is how the interaction *appears* to girls and boys. Once an awareness of a discrepancy is there, the self-consciousness will support experimenting with changes. Most of all, girls may be very supportive of any attempts to change their experience; by sharing the reasons and the methods with pupils, change is much more likely. For example, it might be helpful if, as suggested by Lynn Joffe (1983), girls are made aware of the fact that they – as a group – are less likely to display confidence in mathematics and are more likely to attribute their own failure to lack of ability than the boys. They will probably be quite surprised at this suggestion. One way of drawing attention to this is to discuss with your pupils the results of activity 3.5. (Asking pupils to analyse the questionnaire responses for themselves might provide an appropriate starting point for such a discussion.) Moreover, Lynn Joffe states that girls should be encouraged to highlight areas in which they excel and be discouraged from making statements about their inadequacies in an overly modest manner. Teachers can help here by interacting in appropriate ways and by employing a range of teaching and learning styles which allow all pupils equal opportunities to excel (see section 3.3 here).

3.3 Pupils' preferences for different teaching and learning styles

Conventionally, until recently, mathematics has been seen as a class-taught subject in which information is given by the teacher and the success of its transfer is judged by a sequence of questions from the teacher and answers from pupils. This style of teaching, labelled as 'expository' by Cockcroft (1982), tends to encourage a 'competitive' atmosphere in the maths classroom and – perhaps as a direct result of this – appears to be highly conducive to anxiety, most especially among those whose confidence in the subject is low. Section 2.1 demonstrated that, in general, low levels of confidence are associated with more girls than boys, so a shift away from expository teaching might therefore be expected to support girls (and indeed some boys) in learning mathematics.

What about the use of individualised schemes of work? There is no obvious competitive component here. However, many teachers have found that, although individual workcards *appear* to be non-competitive, nevertheless they can sometimes induce a highly competitive atmosphere in the classroom as pupils compete to complete more cards or to be seen to be further advanced through the scheme. Moreover, individualised work in itself does not necessarily provide pupils with a supportive environment for their learning. Their *perspective* can still be that the mathematics is competitive.

You have no doubt discovered for yourself that pupils respond differently to different styles of teaching, but have you ever considered whether there are any particular implications in this for girls? In 1976, Jim Eggleston *et al.* reported on a study of O-level science classes. They identified three teaching styles:

style 1: 'the problem solvers' – where the initiative was held by the teachers who, by questioning, challenged the pupils to observe, speculate and solve problems;

style 2: 'the informers' – who presented a 'non-practical, fact-acquiring image';

style 3: 'the enquirers' – who used pupil-centred enquiry methods.

They discovered that:

- *style 1* was popular with boys but not so with girls;
- *style 3* was most effective in maintaining girls' liking for science;
- more women teachers used *style 3*;
- nearly half the men teachers used *style 1*.

Several interesting conclusions can be drawn from these results. For example, Jan Harding (1983), in reporting this and other studies, commented: 'It seems that *style 3* – in removing public interaction with the teacher – may enable girls to participate more fully in the class activity, sorting things out for themselves.' Do the results suggest any other hypotheses to you?

Of the three teaching styles identified above it might be argued that both *styles 1* and *2* encourage 'competitive learning'. On the other hand, *style 3* encourages pupils to work together and so supports a more 'collaborative learning' atmosphere in the classroom.

Activity 3.7

This activity invites you to focus on the differences between competitive and collaborative learning and to explore their relevance to your own teaching. It might be easiest to do this activity with a colleague.

 (i) Take a lesson plan, section of a textbook or workcards which you are intending to use and try to decide if the approach is competitive or collaborative.
 (ii) Plan a parallel lesson, but using the other style. For example, if you normally teach using an individualised approach try a lesson in which the pupils work collaboratively in groups.
(iii) Try out both the lesson plans with the same class (or, if you are working with a colleague, you might like to observe each other's classes).
(iv) Canvass opinions from the pupils as to which lesson they preferred and why. Compare notes with your colleague if appropriate. Can you detect any particular learning preferences in different groups of pupils – for example, between girls/boys, confident/anxious pupils, successful/unsuccessful learners? Do your results support the research findings of Jim Eggleston *et al.* and Jan Harding?
(*Note.* You might like to look back at your results to activity 2.2 here since the responses to statements 1 and 3 on the questionnaire should also provide some indication of the learning styles favoured by your pupils.)

Comment

Experience suggests that pupils who are given the option of working together – whether in pairs, threes or fours – do generally prefer this style to any other. During the development of this pack the only class where there was some suggestion that the pupils would rather use an individualised approach was an examination group where the teacher felt that the competition implicit in the examination was affecting the preferences for working.

Certainly in the experiments mentioned earlier in this chapter, where single-sex groupings were formed, there was consistent support from the girls for collaborative, discussion-based learning. Comments here included: 'the atmosphere in the lessons was one of harmony and cooperation and was conducive to hard work' . . . 'none of the girls had any inhibitions about joining in the oral work' . . . 'everyone felt that they were able to obtain a fair share of the teacher's time and attention.'

One teacher wrote: 'This activity made me analyse – or attempt to – the way I teach not only my girls, but the whole class, and it made me far more conscious of the possible effects of a sexist approach.'

It could well be the case that the style of teaching adopted in the classroom might help to explain girls' dislike of mathematics and their under-achievement in the subject. For example, in mixed-sex classes, where boys dominate with a more competitive style of learning, there is some evidence to suggest that girls are affected by a 'fear of success'. It is proposed that being successful in competitive activities – especially those which have a masculine image – poses a girl with a major threat or fear since this implies that she has competed or that she has been aggressive and this type of behaviour is unfeminine. Consequently, girls do not necessarily want to succeed in this type of atmosphere and this could be reflected in their performance. Changing to a more collaborative teaching style could therefore lead to an improvement in girls' performance in mathematics.

Not only are different styles of teaching a factor in the experience pupils have of mathematics; there is now some evidence to support the intuition that teachers have long held that pupils *learn* in very different ways and this too could help to explain girls' under-achievement in the subject. In recording this work and applying it to the learning of mathematics, Rosalinde Scott-Hodgetts (1986) in her article in *Girls into maths can go* reports on two approaches to learning – serialist and holist:

> Gordan Pask and his colleagues have established a strong case for the existence of two distinct learning strategies – serialist and holist; they suggest that the learning performance is regulated by the level of uncertainty at which the learner is prepared to operate. A serialist proceeds from certainty to certainty, learning, remembering and recapitulating a body of information in small, well-defined and sequentially ordered 'parcels' . . . they tend not to look far ahead . . . they are cautious . . . Holists . . . prefer to start in an exploratory way, working first towards an understanding of an overall framework, and then filling in the details; they will tend to speculate about relationships . . . and . . . remember and recall bodies of knowledge in terms of 'higher order relations'.

Rosalinde Scott-Hodgetts goes on to suggest that the successful learners will be those who are versatile (in that they can adopt whichever learning strategy suits the subject matter to be learned). Further, she conjectures that those pupils who are predisposed to learn by a serialist strategy are unlikely to develop into versatile learners of mathematics unless offered a role-model who uses a holistic strategy as well. Her hypothesis is that in the primary mathematics classroom, the (often female) teacher input tends to be serialist and hence encourages serialist tendencies. More girls than boys follow the teacher's lead and tend to adopt an exclusively serialist approach which leads to success – at least in the primary school. Boys, on the other hand, tend to be holists but adopt the serialist strategies offered by the teacher as well. By using a serialist strategy exclusively, it is argued, pupils hamper their *long-term* mathematical development since, although they are better at assessment of familiar content in familiar contexts, they are more thrown by unfamiliar situations than holists. Versatile learners and successful mathematics students will, argues Scott-Hodgetts, have flexibility of choice between different learning strategies. Consequently, boys tend to be more successful long-term.

Hilary Shuard (1986) in *Girls into maths can go* holds similar views about the role of the teacher. She comments on a Schools Council Project in Primary School Mathematics undertaken by Murray Ward (1979) in which 2300 children were given tests in mathematics. The questions reflected 'a list of topics that ten-year-old children might be concerned with'. In some questions the child was told the mathematical operation required (so that only computation was tested); in others the child was required to read the question and to reason out what mathematics was involved. An interesting thing about these test results was the teachers' opinions of the importance of the different questions. The children's teachers were asked how important they thought it was that a ten-year-old child of average ability would be able to answer questions of each type. The questions where significantly more girls were successful than boys tended to be the purely computational ones involving number and money, and these were ranked by the teachers as more important than those involving (say) understanding of place value, measurement, and spatial visualisation – where the boys were significantly better. Hilary Shuard, when commenting on this result, suggests that it demonstrates that the *types* of questions which primary teachers emphasise in class as being important are therefore the straightforward computational ones and that girls seem to respond to this lead, but that the strategies of boys in answering the 'problem' type questions may be advantageous in their later study of mathematics. Furthermore, she concludes her article with the statement: 'it would be mischievous to suggest that pupils who pay attention to teachers' traditional emphasis in primary mathematics give themselves a positive *dis*advantage for future success in mathematics, but evidence seems to point in this direction'. And girls seem to be more likely to follow this path!

Activity 3.8

Consider your own teaching practice in the light of this discussion and try to jot down answers to some or all of the following questions.

  (i) When I learn mathematics myself am I more comfortable working stage by stage (serialist) or do I prefer to explore around a topic, trying to fit it into a total picture and only putting in particular details later (holist)?
 (ii) When I teach mathematics, do I offer more opportunities to serialist or to holist learners?
(iii) In the classroom, do I try to encourage those with one approach predominating to try the other?
 (iv) Are pupils' informal methods actively used and supported in my classroom – thereby validating all approaches – or do I tend to try to encourage my pupils to adopt the method I consider most appropriate?
  (v) Do I raise questions with my pupils about relationships between different aspects of a topic or do I leave it to them to make these connections?
 (vi) Are there gender differences in the way my pupils respond to being asked to take part in investigations?

(vii) Are there gender, cultural or class differences in the topics or mathematical tasks that my pupils find difficult or easy? (You might like to refer back to section 1.4 here.)

Comment

Many primary teachers (often female) concur that their own approach to learning mathematics is serialist and that they feel uncomfortable initially when asked to move from a step-by-step approach to a more speculative, exploratory style. They agree that, in favouring a serialist approach themselves, they tend to provide role-models and so reinforce the same approach in their pupils. This is, of course, exceedingly valuable for any pupils who have developed a holistic learning style themselves, as they will naturally be encouraged towards the very versatility which will stand them in good stead in their later learning; but, as we have shown, it could help to explain why girls under-achieve relative to boys at secondary level.

Secondary-school teachers tend to comment on the serialist approach of most work schemes other than an individualised approach such as the Secondary Mathematics Independent Learning Experience (SMILE) which has a holistic philosophy. Expecting serialistically reinforced pupils to thrive under a holistic philosophy might be too much of a challenge unless the efficiency of the approach is consciously reinforced by the teacher in the classroom to such a degree that pupils are encouraged to become versatile learners. However, many secondary-school teachers have affirmed that their girls are uncomfortable when asked to do investigations and do assert greater preference for secure, clearly ordered learning. This is consistent with the argument that girls have been previously reinforced in a serialist approach. Girls therefore experience heightened anxiety when asked to work in a less structured way.

Do you agree that arithmetic computation – at which girls' performance tends to be better than boys' – is a good example of a task which is performed serialistically? Do you think that a task requiring the visualisation of three-dimensional solids seems to be more amenable to a holistic strategy and so is probably more successfully performed by boys? Can you identify other tasks or topics which can be analysed in this way?

What can be done?

This section has discussed some aspects of teaching and learning styles which seem to have particular implications for pupils' attitudes towards mathematics and also for their mathematical performance. Here we draw your attention to two specific suggestions regarding your teaching approach because these could benefit various groups of pupils – particularly girls.

(1) Try to use an enquiry-based teaching style at least some of the time since this is a way of developing the kind of collaborative learning encouraged in the Cockcroft report (1982). It is implicit in the suggestion made in paragraph 243 in the report that investigations, problem solving, discussion and practical activities should be included in the *full range* of teaching styles in the mathematics classroom. The strength of an enquiry-based method of learning is that it enables pupils to build on the ideas of all those in a group, to confront different interpretations and to try different approaches. By listening to pupils' discussion, teachers can gather new information on what is and is not understood. When pupils work together there is greater possibility that one pupil's understanding

can be shared. However, most of all, the mathematics is no longer public and confrontational, and one pupil or one group can no longer dominate the classroom interactions. Of course, it is possible for an assertive or aggressive pupil to dominate the work of even a small group, and when encouraging collaborative work it is important to ensure that it is not possible for one pupil to dominate in this way. Girls often prefer to work in single-sex groups and this could help to avoid any possible domination by boys.

(2) If versatility in learning style is a way that might ensure greater success in mathematics, then every opportunity should be taken to celebrate differences in approach rather than to reinforce singularity. It could be through pupil collaboration, so that learning styles are shared and the need to choose appropriately for different tasks is underlined. It could be through the interactions of a teacher pointing out these distinctions. It could be through discussion with the pupils themselves, that makes clear the benefits of learning some things by a serialist style and others by exploring the terrain in a holistic manner.

3.4   Summary and reflection

This chapter dealt with aspects of classroom organisation and interaction which have been found to affect the performance of girls (although, because of the different roles assumed by pupils in all classes, the issues raised here should be pertinent whatever your teaching situation).

The first section focused on formal and informal aspects of schooling such as classroom registers, role-models, timetabling, lining up, and so on. While some people do not feel these are very important, they nevertheless represent a way of thinking which emphasises sex differentiation. These are, in a sense, the easiest things about which to do something. The second section looked at aspects of teacher interaction and in particular the use of praise and criticism. We then looked at the implications of adopting different teaching and learning styles in the classroom.

Throughout the chapter we emphasised that the way in which a teacher organises and interacts with her or his class plays a crucial part in the development of pupils' attitudes to mathematics. If there is a relationship (albeit complex) between feelings of confidence or anxiety, and success, then any action that the teacher can take in the classroom in order to increase confidence and to reduce anxiety should help in improving the attitudes of all pupils and could indirectly help in improving their performance in mathematics. Figure 3.4 summarises how the suggestions in this chapter might help in achieving these aims.

Figure 3.4 Suggestions for reducing anxiety in the classroom

Aim	Strategies
Relaxed classroom atmosphere	Reduce competition, avoid authoritarian attitudes, work in small groups, get to know pupils better
The promotion of mathematics as an enjoyable as well as useful activity	Discourage rote learning and promote understanding; use games and puzzles to introduce 'fun' element; give 'hands-on' experience through practical work; emphasise that there may be many ways to reach a solution to a question and/or many possible solutions; encourage guessing and then checking to see whether correct
Clear guidance	Give praise for clear setting out but no extra commendation for tidiness or elaborate artwork; promote a positive attitude to errors; teach good study methods and problem-solving techniques

Activity 3.9

Jot down in the table below the aspects of the problems discussed in this chapter which are most relevant to your pupils or to your school. Beside each of these, note the strategies you feel you can most effectively introduce in order to improve the situation either now or in the future.

Aspect of the problem	Strategies for tackling it

Further reading

■ Articles by L. Burton and R. Townsend, A. Eales, and R. Scott-Hodgetts in *Girls into maths can go*

4 Bias in teaching materials

> The attitudes the reader brings to the printed page affect the strategies he or she uses in processing the message.
> (Zimet 1976)

Language is a very powerful means of communication in our society. From political manifestos to billboard advertisements, from school textbooks to poetry it demonstrates its power to inform, persuade, instruct and entertain. We place great value on literacy and imagery so that we can both express our thoughts and receive the ideas of others. As Sara Zimet implies, the way in which people receive ideas, be they visual, spoken or written, depends upon previous experience. Indeed a study for the National Council of Teachers of English in Illinois by Purves and Beach (1972) found that:

- readers prefer a particular work if the subject matter is related to their personal experience;
- readers get more involved in the material if it is related to them;
- readers seek out words with which they can identify or where there are characters who resemble them;
- readers impute values to characters that are not contained in the story and they often supply their own background to their work;
- when reading material that actually conflicts with their own world view, readers are likely to misinterpret it, to select only the parts with which they agree or to reject the text entirely;
- the more personal or intense the reader feels about the material, the greater is the likelihood that an inaccurate interpretation of the author's intent will be made.

Activity 4.1

The Purves and Beach study was carried out in the USA on written materials for the teaching of English. To what extent do you think that their findings are relevant to teachers of mathematics in the UK?

Comment

You can probably identify a number of parallels. As with any subject, all teachers will agree that written materials are important in the mathematics classroom. To provide pupils with the necessary conditions for mathematics learning they should contain

(a) images to which pupils can relate (in terms of gender, race and class), and
(b) images which are active and successful.

If they do not do so, effectiveness of text as an instrument of learning is reduced. The same is true for other teaching materials used in the maths classroom.

Do the teaching resources used in the mathematics classroom actually assist learning for all pupils? The current trend for putting mathematics in context and making it appear 'friendly', while helping some pupils, could alienate certain groups of children even more if it is done thoughtlessly. Real-life contexts, if they are unfamiliar, are likely to deter pupils. For example, since adolescent boys exert considerable pressure on girls to conform to stereotypes (see section 2.1), it is as important for them as it is for the girls themselves that the way the subject is presented reflects the claim that 'mathematics is indeed for everyone'.

We suggest therefore that authors of textbooks and other materials – such as worksheets – bear considerable responsibility for the way in which the messages they wish to convey are presented and also for the effects and influences that they exert on pupils.

In this chapter we shall look at how males and females are represented in written (and other teaching) materials and the ways that their interests, aspirations and expectations are portrayed. In section 4.1 we investigate the actual representation of people in texts and illustrations, then in section 4.2 we concentrate on the context in which mathematics is set. (Note that the activities in these two sections are aimed primarily at written materials because books, workcards and worksheets are the media most frequently used in schools. However, please do not feel constrained to confine your investigations here to written material; much of what is said applies equally to other teaching materials – such as video, radio, computer software, wall displays and posters.) The third section considers the use of apparatus in the maths classroom. In each of these three sections we identify action that you can take in the classroom in order to improve the situation for girls, although, inevitably, much of this is aimed at *counteracting* the bias displayed in commercially-produced educational materials. In the final section of this chapter we discuss the relationship between teachers and the media, and consider how you might put pressure on publishers to change their practice.

4.1 References and activities portrayed in teaching materials

Activity 4.2

The illustration in figure 4.1 is taken from *Mainstream mathematics*, *Book 1* (Sylvester 1979) at the beginning of a chapter about coordinates. Jot down the messages which you think are implicit for girls and boys in this picture.

Figure 4.1   (Sylvester 1979)

Comment

Even though the illustration is intended to convey a friendly situation taken from real life which will appeal to children aged between thirteen and sixteen and which will provide an example of coordinates being put to a useful purpose, its underlying messages may be very different. For a start, assumptions have been made in portraying all the children in the picture as white. Second, the only girl is sitting down passively, evidently lost, exhausted and nursing her blister. On the other hand, the two boys are actively tackling the problem of where they are by studying the map and relating it to the landscape around them. Third, it is unlikely that pupils of this age-group from an inner city area will ever have been on their own in the country.

So it seems that unless the pupil is a white boy living in or near a rural area (and less than a quarter of the target age-group are in this category) then this illustration serves little positive purpose. It may even actively detract from the mathematics it seeks to enliven.

Activity 4.3

Take a quick look at any worksheet or other materials that you have used recently in class (today if possible), and just count the references to males and females.

Comment

If you are brave you will have looked at a worksheet that you made up yourself. Did your results surprise you at all or were there equal numbers of references to each sex? Did the references reflect a multicultural society?

Of course, it is unfair to take one picture or one worksheet out of context and to criticise it in this way, for it is the overall balance which is important. For example, with figure 4.1 the balance may be redressed in the rest of the book; similarly in activity 4.3 you might reach a different conclusion if you were to examine other materials that you have used recently in the classroom.

Pupils do need to broaden their experience by seeing images of people unlike themselves, but they also need to see images which set the mathematics in the context of the world they know. It is the *balance* that matters, and to achieve this the girls must be seen to be actively taking part in the mathematics and the activities associated with it. This suggests that it is not enough for material simply to refer to equal numbers of girls and boys. Jean Northam (1982) in *Girls into maths can go* draws attention to a number of forms of gender bias which were highlighted by her examination of some mathematics textbooks covering the 5–13 age-range. She records that:

- there are more references to males than females;
- the number of references to girls decreases in textbooks as the target age of the book increases;
- illustrations show girls as lacking individuality, boys and men have distinguishing features;
- famous mathematicians referred to in books are all men;
- the roles that girls play are different from those of boys – boys are depicted as assertive problem-solvers, whereas girls are portrayed in the main cooperating with or helping other people.

Northam's results are supported by statements from teachers who have analysed other textbooks in the same sort of way. A head count of people appearing in a book does generally reveal more male references than female references, particularly in those books aimed at older pupils. Even if the representations of females and males appear to be roughly numerically equal, it is still not safe to assume that the book speaks to the boys and the girls in the same way.

As a demonstration that equal *numbers* of representations of each sex are not sufficient, Jean Northam took a junior textbook, *Maths adventure* (Stanfield and Potworowska 1971), with roughly the same numbers of references to males and females, and categorised the types of activities engaged in by the boys and girls. Her results are shown in figure 4.2. (Note that the categories are not mutually exclusive – some activities fall into a number of categories. Adding up the numbers of references to different activities does not therefore reflect the actual numbers of references as such.)

Figure 4.2   The activities in which girls and boys were engaged in *Maths adventure* (Northam 1982)

	Activity	Girls	Boys
(1)	Identifying, setting and solving problems	10	27
(2)	Teaching or explaining processes to others	10	35
(3)	Making something or displaying a skill	2	9
(4)	Planning, initiating or inventing	3	11
(5)	Performing, playing tricks or boasting	0	12
(6)	Competing	2	7
(7)	Repeating or elaborating on a process already learnt	19	8
(8)	Cooperating, sharing, helping or complying	12	6
(9)	Correcting another's behaviour	13	3

A closer examination of the way children and adults are portrayed may show differences in behaviour which mimic stereotyped attitudes towards the sexes and perhaps to race and class. As figure 4.2 indicates, the boys in *Maths adventure* are engaged in activities which require initiative, prowess and performance, while the girls are mainly taking supportive, non-competitive roles. What is more, during the development of this pack, one teacher looking at the same book found that the actual sizes of the illustrations of boys were bigger than the illustrations of girls; i.e. that the boys took up more physical space on the page.

Activity 4.4

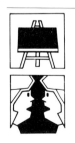

This activity invites you to analyse in some detail a textbook or scheme of work for gender bias.
 (i) Duplicate three or four copies of the chart in figure 4.3. Record on these all the references to people which occur in a mathematics textbook or scheme of work that you use, and describe briefly what those people are doing. (If it is convenient, you might like to work with a colleague or pupils here.)
 (ii) Categorise, using tallies on the table in figure 4.4, the behaviour of all the people referred to in your selected textbook or workscheme.
(iii) Discuss with colleagues the kinds of bias that you found in the materials that you examined.

Figure 4.3 References to males and females in a textbook or workscheme

Book/workscheme _____

Publication date _____   Target age/level _____   Age of pupils doing the work _____

Topic	Page or card number	Context: text, exercise, illustration	Reference and name if given	Description of activity e.g. Setting/solving problems, passive/active, asking questions, earning money, being in charge, being successful, collaborating, obeying instructions, etc.

Figure 4.4   Types of activities engaged in by males and females in the textbook or workscheme

	Activity	Girls	Women	Boys	Men
(1)	Identifying, setting and solving problems				
(2)	Teaching or explaining processes to others				
(3)	Making something or displaying a skill				
(4)	Planning, initiating or inventing				
(5)	Performing, playing tricks or boasting				
(6)	Competing				
(7)	Repeating or elaborating on a process already learnt				
(8)	Cooperating, sharing, helping or complying				
(9)	Correcting another's behaviour				
Other					

Notes

(a) The categories used in figure 4.4 are the same as those employed by Jean Northam in figure 4.2. You may like to compare your results with hers. If you find that the categories are not quite appropriate for you, or if there are not enough categories, please feel free to devise additional ones.

(b) You might also consider going through the text and the illustrations separately rather than putting them together since it has been suggested that this is likely to provide a better indication of the overall balance in the materials.

(c) You could also use figure 4.3 to analyse a textbook or workscheme in terms of race and class bias.

Comment

When teachers carried out this activity at workshops they were surprised at just how much bias they found in the materials they looked at. They were also shocked that newly published materials did not seem to be much of an improvement on older ones. Figure 4.5 indicates the results obtained by one group.

Quite apart from the contexts in which the mathematics was set (which will be considered separately in the next section) teachers reported:

- fewer female references overall;
- very few adult women;
- adult women often portrayed in stereotyped roles, such as in the home or shopping – or in a subsidiary role as someone's wife, mother, aunt, etc.;
- adult men portrayed doing more activities outside the home, such as earning money;
- many girls (and few boys) behaving in passive ways, (say) sitting or standing watching something;
- many boys (and few girls) behaving in active ways, for example climbing things;
- names used to imply stereotypes, for example Crafty Kate, Basil Brayne;

Figure 4.5 One group's findings

Book/workscheme A Natural Approach to Mathematics (Part 3)

Publication date 1970 ___ Target age/level 12/13 ___ Age of pupils doing the work Not used by a class.

Topic	Page or card number	Context: text, exercise, illustration	Reference and name if given	Description of activity e.g. Setting/solving problems, passive/active, asking questions, earning money, being in charge, being successful, collaborating, obeying instructions, etc.
Fractions	24	Exercise	John	– buying sweets. – Active.
"	24	"	Christine	– reading a comic on a train. – Passive.
Percentages	36	"	Achy	obtaining 65% in a test – Being successful.
Money	37	"	Shopkeeper	buying & selling lady's coat – Being successful, earning money.
Arithmetic	39	"	Boys	gaining points in sports – Active, being successful.
Algebra	41	"	Jane & Mary	saving & spending pocket money.
Measurement	46	Picture at beginning of text.	Two men (one English, one French)	with Imperial & Metric weights respectively.
Coordinates	55	Text	Mr Smith	'Mr Smith's ticket was (3,7). How would you expect Mrs Smith's ticket to be marked.' Women being passive.
Coordinates	57	Exercise	Angela	'Where could Angela sit to be certain of having boys in the desks immediately in front of her & behind her.'
Statistics	64	Picture at beginning of chapter.		Shows a male statistician and a boy collecting data. – Solving problems, being in charge.
Algebra	83	Exercise	A boy	A boy marks his own work – then he loses marks for untidiness & incorrect spelling.
Percentages	86	Exercise	Tom	Tom's marks in a maths test.
Conversion Graphs	109	Exercise	John	What is the ratio of John's earnings to the total earnings of the whole family. – John as breadwinner.
Averages	125	Picture at beginning of chapter.		Picture giving idea of average height but all people are boys or men.
Arithmetic (Eq'ns)	175	Exercise	Sarah	A class was told to halve their marks and add 9. Silly Sarah was not listening.

- more male pronouns even when no specific names were mentioned;
- nearly all people were white.

They also noticed that secondary-school maths texts are written mainly by men! Were your findings similar?

What can be done?

It is unrealistic to try to eradicate all forms of bias from educational materials at once. Teachers cannot expect to restock their mathematics departments completely – even if materials existed which avoid all the pitfalls identified above. But there is action that you can take.

(1) When you are preparing your own material you can monitor the balance of women, girls, boys and men; you can make sure that there are roughly equal representations of the sexes and that the behaviour exhibited by women and girls is as dynamic, active and interesting as that of the men and boys. Being aware of how bias is portrayed should help you with this.

   You can also attempt to counteract the impression conveyed by any commercially produced materials that you use.

(2) Because mathematics is approached nowadays in different ways – with more emphasis on practical work, pupil enquiry and collaboration – there are many opportunities for classroom discussion. The assumptions inherent in visual, spoken and written images provide valuable starting points for such discussions about the roles of men and women, girls and boys, and people from different cultural or class backgrounds. Say to your class that whenever you come across a stereotyped image you will draw their attention to it and invite them to do the same.

(3) Make a point of reversing all gender references in the next piece of text that you use in class. Alter all 'she's' to 'he's' and vice versa. Explore your own, your colleagues' and your pupils' reactions. You may be surprised at the way in which this can alter your interpretation of what is written. For example, consider the extract in figure 4.6 taken from *A way with maths* (Langdon and Snape 1984). Do you usually assume that prisoners are men? Many such attitudes are deeply ingrained. In order to challenge these it is important that they be explored and discussed. Reversing gender

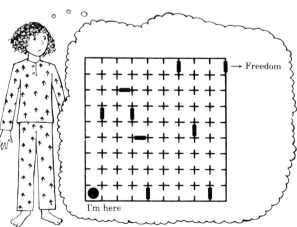

Figure 4.6 (Langdon and Snape 1984)

The prisoner's electronic nightmare

A prisoner was held in a modern electronically controlled prison which had 145 doors. But something strange went wrong with the control and most of the doors were left open. Nine of the doors, shown in black, were locked but each one would open if, before she reached it, the prisoner had passed through exactly eight open doors. However, as the prisoner passes through a door, it automatically locks behind her.

   The prisoner, in the bottom left-hand cell, had a plan of the prison. She thought carefully for a long time and then walked through all nine locked doors to freedom. *What was her route?*

differences could encourage people to recognise attitudes that they do not realise they hold.

(4) More generally, the use of sexist language and its influence on social behaviour has been discussed widely over the past few years (see Spender 1980) and could provide a useful topic for class and staffroom discussion.

Of course, these suggestions do not resolve all the problems. For example, you might like to discuss with your pupils whether life should be shown as one might like it to be rather than the way it is.

4.2   Contexts in which mathematics is set

The Cockcroft report (1982) includes the following statement:

> the applications of mathematics which are found in many textbooks and examination questions reflect activities associated with men more often than with women,

thus confirming that in many ways mathematics can still be classified as a boys' vocational subject – in spite of changes in the curriculum (see sections 1.2 and 1.4).

Nonetheless, the statement draws attention to a difficult dilemma faced by anyone writing teaching materials. It is desirable to set mathematics in contexts which make it meaningful for pupils and which motivate them, and it is sensible to use everyday situations to do this. For example, mathematics material aimed at poorly-motivated fourth- and fifth-year pupils often places mathematics in the home or the workshop. However, it is important that these situations are as relevant to girls as they are to boys, whatever their cultural or class background. It is equally important, for instance, that shopping is not *only* associated with girls nor working with equipment *only* with boys in order to avoid perpetuating stereotypes that already exist.

Activity 4.5

Check through one of the mathematics textbooks or workschemes that you are currently using. How often do the applications suggest male activities even if men or boys are not specifically portrayed?

Comment

Did you find frequent references to ships entering and leaving ports, racing cars, petrol, ladders and cranes etc. And did you find occasional references to items such as flower-pots, haberdashery and so on? Do you agree that such applications reinforce the idea that a lot of mathematics is not really aimed at girls?

We suggest that this activity raises three issues. First, the contexts used often imply unnecessarily that girls and boys have different interests. Second, the so-called boys' interests are catered for more than those of girls. Third, the context often conveys the impression that boys' interests are superior or more important. We now discuss (briefly) each of these in turn.

Implication of different interests

This is highlighted by the illustration in figure 4.7, taken from the School Mathematics Project (SMP) lettered series, *Book F*. It is quite unnecessary and irrelevant to differentiate by gender here; it cannot be assumed that all girls knit any more than it can be assumed that all boys mix concrete. The idea is

Figure 4.7   (SMP *Book F* 1970)

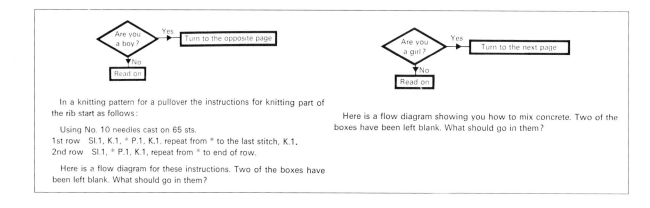

In a knitting pattern for a pullover the instructions for knitting part of the rib start as follows:

Using No. 10 needles cast on 65 sts.
1st row   Sl.1, K.1, * P.1, K.1, repeat from * to the last stitch, K.1.
2nd row   Sl.1, * P.1, K.1, repeat from * to end of row.

Here is a flow diagram for these instructions. Two of the boxes have been left blank. What should go in them?

Here is a flow diagram showing you how to mix concrete. Two of the boxes have been left blank. What should go in them?

simply to familiarise pupils with the techniques involved in working through a flow chart and that is why the yes/no question is introduced at the beginning. The question 'Are you a boy?' implies that the examples that follow about knitting and concrete are gender-specific, and they need not be since they do not actually relate to the initial question. However, a relationship has now been established in the reader's mind which connects gender with particular activities. Why did the authors not ask simply 'Do you like knitting?' Can you identify similar differentiation of interests in the materials that you use?

Preponderance of supposed boys' interests

Jean Northam (1982) in *Girls into maths can go* draws attention to the dominance of supposed boys' interests at junior level, when she writes:

> Maths in the junior books moves away from the domestic sphere and takes to the sports field, the battleground and the worlds of business, space travel and machines where young women are not to be found . . . the women are almost without exception confined to sitting in cars and buses driven by men at work, or being rescued by firemen . . .

The situation seems to be no better – or even worse – at secondary level. For example, Clwyd County Council (1983) carried out a survey of option selection guidance materials for secondary pupils. These materials were used to assess the facility with which children might follow a course of study to examination level and their suitability for specific courses. The study revealed that the vast majority of mathematics questions mentioned males or male interests.

You will have seen from your own investigation in activity 4.5 the extent to which this is true in the materials that you use.

Assumption that boys' interests are more important

The assumed importance of the activities carried out by men is often related to the world of work, finance and sport. Clwyd County Council (1983) specifically mentions the following contexts:

'Mr Jones is buying a house . . .'
'If Mr Davies borrows . . .'

'The employee Gareth . . .'
'Mary's starting salary was £1100 per year . . .'
'The goals scored by a football team . . .'

and it reports:

> The implication is that men fill the world of work and finance. The only questions with a reference to a woman placed her very low down on the earning scale.

As well as forming part of the hidden curriculum for girls, this bias accounts at least in part for the conviction of many adults from all backgrounds that mathematics is not only impossible to understand, but is also largely concerned with business, tax, accounts and insurance policies. In other words it is seen as being an important tool which people by and large do not consider themselves competent to use. There are also other applications of mathematics which are usually associated with men, for instance applications to navigation and weapons. Such associations tend to crop up more frequently in course material for older and more highly achieving pupils; might this also reflect a presumed higher status of male interest?

We consider now in more detail the effect that the context of the mathematics might have on pupils. In the USA, Richard Graf and Jeanne Riddell (1972) asked a small sample of male and female students to attempt one of the problems in figure 4.8. The mathematics involved in these problems is identical; the only difference is in the contextual setting.

Figure 4.8   (Graf and Riddell 1972)

At a fabric store, satin costs $3.50 per yard for the first two yards and $3.25 for each additional yard. Lace costs $1.30 for the first yard and $1.10 for each additional yard. How much would it cost for the material to make a dress requiring four yards of satin and five yards of lace?

A stock broker charges the following commission for his services. For making purchases, he charges $3.50 per purchase for the first two purchases and $3.25 for each additional purchase. For selling stock, he charges $1.30 for the first sale and $1.10 for each additional sale. How much would he charge for making four purchases and five sales?

Richard Graf and Jeanne Riddell found that the female students doing the stockbroker problem took longer than the male students doing the same problem. The female students doing the fabric store problem took the same time as the male students who did it. There was no difference in accuracy between the girls and the boys on either problem. The researchers infer that the stockbroker context is male and the fabric store context is female and that this difference in context accounts for the different times taken by the sexes to reach a correct solution of the 'male' context problem. This may have been so at the time and place of that experiment; or at least the fabric store problem might have been seen by the students as just 'less male' than the other one. However, when a British teacher tried out these two questions on her mixed fifth-year O-level class, the pupils pointed out that *both* the questions were unrealistic for them. Material is not sold at one price for the first length with a different price for subsequent lengths, and stockbroking is not a familiar occupation.

On the other hand, other studies in Britain confirm that the context of a question does seem to be important. Not only do girls seem to be slower than boys in tackling questions framed in a male context, but they may also not perform so well even if time is not a factor. For instance, in a study of 1750 second-year pupils in Sheffield comprehensive schools (Eddowes, Sturgeon and Coates 1980), girls performed better on a question referring to the area of dress material than to an equivalent one referring to the area of metal needed for a template, and they also performed better on a question asking them to calculate quantities required for a recipe than on a similar question relating to a blast furnace. Boys seemed to do equally well on both.

Activity 4.6

This activity invites you to investigate whether your pupils have any preferences for particular contexts.

 (i) Duplicate the investigations in figures 4.9 and 4.10 and ask pupils in one of your classes to choose which one they prefer to do. Record how many pupils choose each investigation.
 (ii) After they have finished working on the investigations ask the pupils what directed their choice.

Comment

Did your pupils choose one investigation because it appealed to them or because the other looked worse? As with the Graf and Riddell problems, the mathematics in both investigations is virtually the same, but it is presented in different contexts. Did more boys choose investigation B because it is set in a male context, and were the girls put off by this male context? Were there any other reasons for pupils' choice? For example, investigation A has much more written language in it and this could be off-putting to some pupils. Did any pupils in your class avoid investigation A for this reason? Your pupils may also have other views about what does and does not appeal to them, which you may like to discuss. This could help you in devising your own material in the future.

What can be done?

The discussion in this section has identified three possible problems for girls in connection with the context in which their mathematics is set: they might be 'put off' by a problem or a piece of work set in a masculine context, they seem to take longer than boys to complete questions set in such a context, and there is also quite a bit of evidence to show that girls do not perform as well as boys on questions set in a masculine context. (This point is picked up again in chapter 5.)

There are two ways in which you might like to tackle these problems when devising your own materials.

(1) If there is a context at all it should – ideally – be meaningful to all pupils. If the contexts are tied to specifically male activities then they will not necessarily be meaningful to girls and so could adversely affect their performance. It is all the more important therefore to avoid any unnecessary stereotyping which might reduce the activities in which girls (and some boys) feel they can wholeheartedly participate. On the other

Figure 4.9 Investigation A

You will need a piece of A5 paper and some scissors for this investigation.

You run your own newspaper. Customers pay to advertise in your paper and the price they pay depends on the size of their advertisement.

You need to choose some advertisements to fit onto your piece of paper. Unfortunately, you will not be able to use all the advertisements.

The price the customers pay is written by the sides of the advertisements. The more space that you can fill up, the more money your newspaper will take.

What advertisements will you choose?
How will you place them?
How much money will you take?

David's Dancing Gear

Catsuits
Headbands
Ribbons
Ideal Gifts
Selection of
dance shoes
Tights
Make-up
Accessories
Sweat Pants

£290

£75

HAYWARD GALLERY

SOUTH BANK, LONDON

DEGAS
THE PAINTER AS PRINTMAKER

also showing THE HAYWARD ANNUAL:
RECENT BRITISH ART

Until 7 July

ARTS COUNCIL

Devon Truck-Hire

SELF-DRIVE
VANS, TRUCKS
AND CARS
9 cwt. to 4 ton

Competitive rates for short or long rentals and journeys. Check our rates first — our special cheap rates will save you even more. It's worth a telephone call! And now tail-lifts available on box vans.

£75

LADIES &
GENTLEMEN

If there was a way for you to make an extra £500 to £5,000 plus per month over and above your present income, would you be interested?

 **ANGLIA FASHIONS**

requires a

JUNIOR SECRETARY

Applicants, aged 18-22, should have "O" level English and Maths, first-class shorthand-typing skills and preferably some commercial experience. Good appearance, speech and confident telephone manner also required.

We offer a competitive salary, dress allowance, subsidised lunches, etc.

£120

£75

 £110

£180

Jenny Craig Weight–Loss Centre

LOSE WEIGHT

One month slimming course, with Herbal Aide, is absolutely free.
If you are content being overweight and unhealthy, fine, if not...

The Italian Suit Specialists from £89·50
The Italian Suit Shop

91

Figure 4.10 Investigation B

You will need a piece of A5 paper and some scissors for this investigation.

You own a DIY store. Customers are in your shop wanting to buy particular pieces of chipboard. Scale drawings of the shapes and sizes that they want are shown below. Your sheet of paper is a scale drawing of a large piece of chipboard from which you will cut the pieces for customers. Use the drawings to help you decide how you would cut up the chipboard in order to create as little waste as possible. (Unfortunately, you will not be able to satisfy all your customers.)
The prices at which you will sell the wood are marked on the pieces.

What is the most money that you can take?

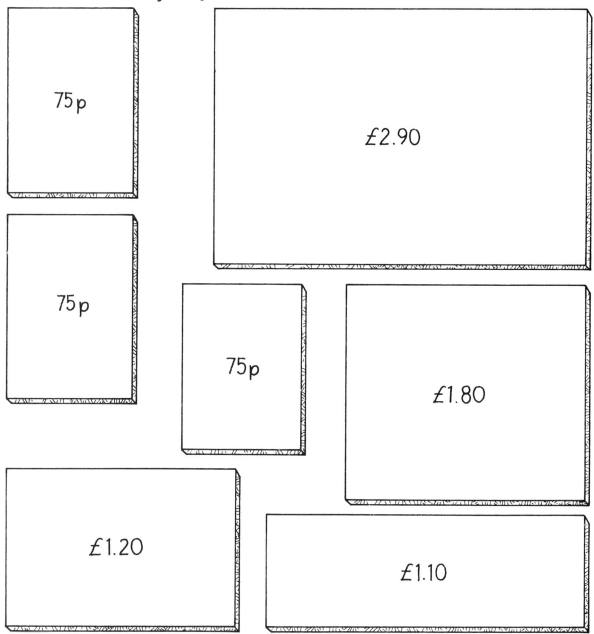

hand, it is also very important to provide all pupils with work which they can relate to and which helps them to make mathematics 'their own'. Activity 4.6 may help you to identify meaningful contexts for all your pupils.

(2) Another way of avoiding stereotyping is to keep the problems 'people-free'. Unfortunately, people are sometimes so worried about being sexist or racist that they dehumanise material completely. This need not be the case; for instance, the milkcrate example in figure 4.11 taken from *Investigator 1* (SMILE 1984) is eyecatching, interesting and full of mathematics.

(3) You can also draw pupils' attention to bias in commercially produced materials and discuss with them how it might be remedied.

Figure 4.11   (SMILE 1984)

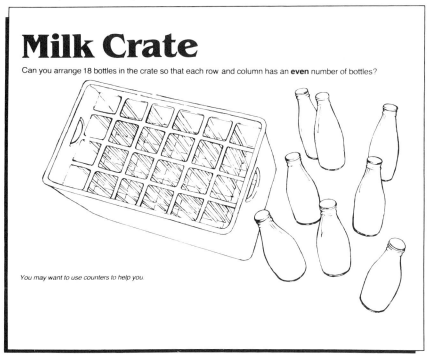

Milk Crate

Can you arrange 18 bottles in the crate so that each row and column has an **even** number of bottles?

You may want to use counters to help you.

4.3   The use of apparatus in the maths classroom

Confidence and assurance in handling mathematical apparatus was found, by the Assessment and Performance Unit (APU 1981), to be the province of boys rather more than girls. It seems that the use of equipment – which is intended to make mathematics more enjoyable and more relevant for everyone – could again reinforce the view that maths is not for girls.

Activity 4.7

(i) Jot down the types of apparatus that pupils used in your maths lessons last week.

(ii) Can you identify any difficulties experienced more by girls than by boys?

(iii) Do you think that you have different expectations of girls' and boys' capabilities in handling equipment?

Comment

The biggest impact on school mathematics recently has surely been caused by the introduction of both the calculator and the computer within a relatively short space of time. These are being increasingly used in mathematics classrooms; both have affected the philosophy of maths education and both are beginning to have a radical effect on the maths curriculum. The calculator is revolutionising the teaching of arithmetic; the need to learn many algorithms is disappearing, leaving teachers to ponder exactly *what* arithmetic skills are most useful to pupils, which ones they should be able to *understand* and which they should be able to *apply* with or without a calculator. The microcomputer is also beginning to make its presence felt in the maths classroom, raising important issues which are discussed in the appendix of this pack.

The extent to which other apparatus is used, and also its nature, is determined largely by the age of the pupils. Toys and games are used frequently by all pupils at junior level, partly to aid the development of coordination skills and partly to help explain fundamental mathematical concepts. At secondary level the emphasis gradually shifts towards the use of 'tools' and 'instruments' in order to aid specific tasks.

We hope you found that girls and boys have equal access to apparatus in your classroom, although one problem experienced by many girls is that they may not be very familiar with it because of their limited access to mathematical toys and equipment outside school. For instance, in section 1.3 you saw that parents often provide different types of toys for their sons and daughters. This not only affects very young children; it also seems to be the case that whilst calculators are becoming more common – to the extent that they are now being given away as free gifts with (say) petrol – fewer girls than boys actually have one of their own; many pupils have even less access to home computers.

Activity 4.8

What is the situation in your school?

  (i) Give the mini-questionnaire in figure 4.12 to the pupils in one of your maths classes.
 (ii) Compare the results according to the kind of family: girls in girls-only families, girls in mixed families, boys in mixed families, boys in boys-only families. Are the results a matter of concern to you?

(*Note*. This activity is adapted from one in Straker (1986).)

Comment

Anita Straker (1986) in her article in *Girls into maths can go* comments on a study that she carried out with primary school children throughout the country in early 1985. The results indicate that boys are almost twice as likely as girls to have their own calculator, to have a microcomputer at home and to possess a digital watch. Furthermore, girls from 'girls-only families' are the least likely to have access to this technology out of school. Were your results similar? Can you identify any particular problems for pupils from some cultural or class backgrounds?

Eventually the calculator will become so common that it will no longer be viewed as a piece of technological gadgetry but as an essential tool which is used with as much confidence and frequency as the ball-point pen. Until that

Figure 4.12 Do you have the following?

1.	Do you have a calculator of your own?	Yes/No	☐
2.	Do you have a calculator in your family?	Yes/No	☐
3.	Do you own a digital watch?	Yes/No	☐
4.	Do you have a microcomputer in your family?	Yes/No	☐
5.	Do you have any brothers?	Yes/No	☐
6.	Do you have any sisters?	Yes/No	☐

7. Please tick Girl ☐ Boy ☐

day the results of Anita Straker's study must cause some concern. Familiarity with these tools at home must increase the confidence with which they are used at school and the range of activities to which they can be applied.

As to whether teachers have stereotyped expectations, consider the following extract which suggests that, as with parents, teachers can influence pupils' attitudes by not providing appropriate encouragement.

> There was no perceptible difference in the structures made by the boys and the girls. Nor was there any difference in their ability to manipulate the materials . . . despite the fact that the teachers themselves thought that the girls would not play with Lego, when this was suggested the girls could and would do so quite effectively. We would suggest (with writers such as Byrne, 1978) that it is lack of opportunity and encouragement which leads to the situation in which girls do not play with construction toys . . . Further, on the occasions when this does happen, teacher expectations are such that the activity may well go unnoticed.
>
> (Walden and Walkerdine 1982)

How relevant do you think that this comment on observations of children at nursery school is to the antipathy of older girls towards mathematics and to the decline in their performance through the years of schooling? How much are the apparent expectations of some nursery teachers paralleled in the expectations of older pupils? If what Rosie Walden and Valerie Walkerdine observe at nursery level persists throughout schooling, i.e. that teachers do not *expect* girls to use apparatus, then there are very serious implications for girls, particularly in connection with the introduction of important new technology into the classroom. For example, there are extremely interesting developments in Computer Aided Learning related to mathematics. One of these is LOGO, a programming language in which pupils use graphics to learn programming skills, explore space and solve problems that they come across whilst doing so. As children are working on their own self-defined investigation with the micro, they themselves decide on the direction that their investigation takes. The problems that arise are therefore particular to their own work. Both girls and boys enjoy LOGO, and are motivated to think mathematically without the contrivance of a 'real life' situation in which to place the mathematics – the 'real life' is the pupils actually sitting at the machine. It is therefore important that all pupils are encouraged to become familiar and confident with the microcomputer.

What can be done?

Two studies confirm that girls lack encouragement rather than competence in handling mathematical apparatus, and these suggest possible ways forward.

(1) Lisa Serbin (1978) provides some evidence that young children are more likely to play with 'opposite sex' toys if there is a teacher of the opposite sex nearby, and that girls are more likely than boys to do this. So if you are a female teacher in a primary school, you might like to ask a male colleague – or a child's father – to assist in your classroom occasionally and to observe the effects. You can also actively encourage children to play with 'opposite sex' toys by making appropriate equipment available to them and by making suitable suggestions.

(2) Barry Everley (1981) describes a technology programme in a secondary school during which girls were exposed to non-typical materials such as Meccano. Girls and boys were asked to construct the same items (a gradient finder, a minimum-focus-tester, an anemometer, a thumb-muscle tester, a motor-power tester and an electromagnetic reaction timer) using nuts, bolts, axles and bearings. The girls showed generally more capability and greater productivity than the boys, and just as much interest. You might like to try something similar in your classroom.

(3) Finally, in this section we have mentioned only very briefly the problems experienced by girls in connection with microcomputers: girls are simply not using the machines as much as boys, and they are less likely to own a home computer. These issues are discussed in more detail in the appendix of this pack in which the problems experienced by girls in all aspects of computing are investigated. There, we suggest a number of ways that teachers can improve the situation for girls – both by trying to improve girls' attitudes towards computing in general, and by creating an environment in the mathematics classroom where there is time and space available for all pupils to use the microcomputer with confidence and assurance.

4.4 The relationship between teachers and the media

We considered earlier in this chapter the type of action that teachers can take in the classroom in order to redress the balance and to play down the occasional downright prejudice displayed in commercially published materials. We now investigate what can be done to prevent the appearance of such biased materials in the first place.

Newer resources may be responding to the changing approach to mathematics, yet there is still a body of opinion which reflects complacently that the world is as it is and that justifies the use of stereotyped images in material in terms of increased interest and involvement (although as you have seen, this may be aimed at one group of pupils at the expense of others). The correspondence in figure 4.13 between a television viewer and a well-known cereal manufacturer illustrates this viewpoint.

Even when authors are aware of such concerns there is sometimes little that they can do in practice. For instance, Laurie Buxton (1984) wrote a book as part of the 'Everyman' series, entitled *Mathematics for everyman*. In America this work has been retitled *Mathematics for everyone*, reflecting the concern of the author. Does the change in title alter your expectations of the book? We

Figure 4.13   Correspondence between a television viewer and a cereal manufacturer (GAMMA 1984)

Dear Sir/Madam,

As a mathematics teacher for twelve years now doing research into why girls under-achieve in mathematics, I feel I must protest at the content of your most recent T.V. advertisement for 'Rice Krispies'.

To infer, even in a humorous manner, that girls who are good at mathematics are in some way 'weird' 'funny looking' or 'unfeminine' does both yourselves and 50% of your customers a great disservice.

There is no research to date which has found any correlation between mathematical ability and unfeminine behaviour or attitudes in girls. The only relevant research has shown that the girls who persevere with mathematical studies, against the pressures which you reinforce with your advertisement, are likely to be as successful (in fact slightly more so), than their male counterparts.

You are well aware of the effects of advertising on young children and I feel that you carry a great responsibility not to reinforce predjudices and ignorance. Mathematics teachers have for many years been trying to encourage girls to make the most of their mathematical talents and we would prefer to have you as an ally in this difficult task.

I feel sure that with the talent available to you, you would be able to devise advertisements which would sell 'Rice Krispies' <u>and</u> encourage all children to make the most of their talents and abilities. In the long term it would be of benefit to us all.

Yours faithfully,

Dear Mr. Allder,

Thank you for your letter of 25th November, concerning our RICE KRISPIES commercial.

We do appreciate the points you make but despite this, our research and consumer correspondence indicate that the majority of viewers accept this advertisement for what it is - a humerous vignette which accurately reflects the conversation and attitudes of boys of that age.

Our advertisements are thoroughly tested and researched amongst housewives of all ages and backgrounds before they are screened, and we take great care to ensure that they represent what our consumer see as 'real life'. We took particular care with this series of commercials to probe for negative views and found very few, so we are satisfied that they are acceptable both ethically and commercially. Finally, you may be surprised to learn that the advertisements were created and written by a lady!

I hope that whilst you may not agree with the advertisement, you now have an understanding of why it will continue to be shown.

Yours sincerely,

suggest that what is possible in the USA is surely also possible in the country of origin of the 'Everyman' series!

An even more extreme example of the limited influence exerted by authors was quoted by a teacher during the development of this pack. The author of a new computing book quite deliberately illustrated the textbook with people in non-traditional roles; there were pictures of women using micro computers and so on. When the book went to print the graphic artist responsible for drawing the illustrations simply assumed a mistake had been made, or maybe did not even notice; at any rate, all the pictures were altered so that they reflected the traditional roles avoided by the author – without even referring back to the author to check! Anecdotes like this are legion. So it is not good enough for some authors to be aware of concerns regarding bias. Publishers need to be aware too. Admittedly, publishers are becoming more sensitive to the concerns of teachers regarding stereotyping in mathematics textbooks, and they are often receptive to suggestions as to what can be done. For example, in recent years, the Gender and Mathematics Association group (GAMMA) has written – with some success – to many publishers, drawing their attention to the ways in which bias can be displayed and asking them to avoid unnecessary stereotyping.

Publishers can also be helped to a better understanding of what is required by being invited to conferences and meetings in order to talk with teachers. For instance, a sympathetic publishers' representative attending a 'Girls Mathematics Day' held recently at a local school was shocked when one of the books he proudly displayed was criticised by teachers for the illustrations it carried. The company believed the book to be an example of *good* practice for there were indeed pictures of women – one was lying on a beach somewhere, one was an actress in a three-part radio play and one was on a board of directors. There were several men on this particular page; apart from the two other actors and the rest of the board of directors there was an airline pilot, a photographer and a television reporter. The teachers explained to the bemused representative that token references to women do more harm than good. In any case the scenes reflected a white, middle-class set of values that were unlikely to do more than exasperate many of the pupils for whom the book was intended.

What can be done?

Some publishers already have policies which aim to avoid stereotyping, and if some can do this then surely so can the rest. Whether publishers are or are not aware and sympathetic, they might take care to avoid images which discourage girls (and some boys) if enough pressure is put on them. We suggest that it is possible for teachers to draw the attention of publishers to practices which may contribute to the under-achievement of girls in mathematics by objecting to materials already in print.

(1) You can write to publishers requesting that they use guidelines (aimed at avoiding bias) *before* publication. Figure 4.14 provides one such set of guidelines to which you may like to refer if you wish to devise your own letter. Alternatively, the Gender and Mathematics Association (GAMMA) have produced a standard letter to publishers which you could use. (The additional information at the end of the chapter gives an address from which this can be obtained.)

Figure 4.14 A publisher's checklist for analysing gender bias (Cambridge Educational)

Checklist for analysing gender bias

YES   NO

___ ___ 1 Are equal numbers of girls and boys portrayed in the text, illustrations and examples?

___ ___ 2 Are equal numbers of women and men portrayed in the text, illustrations and examples?

___ ___ 3 Is the male noun or pronoun (man, he) used to refer to all people?

___ ___ 4 Are women shown only in stereotyped or subsidiary roles, for example as housewives, nurses and secretaries, or as someone's wife, mother etc?

___ ___ 5 Are men shown taking an active and competent part in housekeeping and childrearing?

___ ___ 6 Do females and males participate equally in physical and practical activities?

___ ___ 7 Do females and males participate equally in science-based and technical activities?

___ ___ 8 Do females and males participate equally in arts-based and domestic activities?

___ ___ 9 Are the situations depicted equally within the experience of girls and boys?

___ ___ 10 Do the situations depicted have equal interest for girls and boys?

___ ___ 11 Are females portrayed in more passive roles (for example, sitting, watching) and males in more active roles?

___ ___ 12 Are all careers and option choices portrayed as equally possible for girls and boys?

___ ___ 13 Are females and males portrayed as having equal status at work and at home?

___ ___ 14 Are females and males presented as being equally competent in both intellectual and practical activities?

___ ___ 15 Are females and males described in stereotyped ways (for example, females sensitive and males aggressive)?

(2) You can make it clear to publishers that stock cupboards are piled high enough with poor, badly written, inappropriate mathematics textbooks ordered in previous years and that you object to paying for new ones which do not provide suitable models for girls as active and successful doers of mathematics and capable independent members of adult society.

(3) You can critically read new books from the point of view of equal opportunities and where possible write reviews to appear in widely-read newspapers and journals.

4.5   Summary and reflection

The messages conveyed in this chapter may be summarised by the following quote from the Cockcroft report (1982).

> Authors, publishers and teachers should ensure that written material in mathematics does not reinforce the stereotyping of boys as active exploratory problem solvers while girls are portrayed as passive helpers whose interests do not extend beyond fashion and the home. Applications of mathematics should encompass those with which girls as well as boys can identify. Teachers also need to ensure that mathematics itself is not presented as a male-domain in the daily oral work of the classroom as well as in the written materials.

We have shown that too often mathematics is geared towards successful men, but mathematics should be a socially relevant subject for *all* groups of society. It is up to teachers, in the first instance, to demonstrate to pupils that this is indeed so, and to provide them with appropriate materials and opportunities in school.

Activity 4.9

Jot down in the table below the aspects of the problems discussed in this chapter which are most relevant to your pupils or to your school. Beside each of these, note the strategies you feel you can most effectively introduce in order to improve the situation either now or in the future.

Aspect of the problem	Strategies for tackling it

Further reading and additional information

- Articles by J. Northam and A. Straker in *Girls into maths can go*
- The Gender and Mathematics Association (GAMMA) can be contacted through
  Marion Kimberley, Department of Mathematical Science, Goldsmith's College, New Cross, London SE14 6NW

5 Assessment

I was always slow at maths tests. I couldn't leave a question until I thought I'd got it right. I can remember one test when I got an answer of a half. It was the price of something. I puzzled over it for ages as I knew it couldn't be half a penny – nothing was that cheap. I checked and rechecked my working and then I realised that I'd worked in pounds so it really was 50p and my answer was right all along, but I'd spent so long on the one question that I didn't have time to do much more.

(Judith, aged 18)

The most obvious and widely used indication of mathematical achievement is provided by pupils' performance in examinations and other tests. Indeed, DES statistics on achievement in public examinations (see figure 0.1) formed one of the starting points for this pack since they highlighted (amongst other things) the fact that on average girls are not so successful as boys in public examinations at 16+.

Of course, public examinations, whilst being important for girls' careers, only give a picture of the situation at 16+. The problem of girls not doing themselves justice does not suddenly start at 16. It could go right back to the primary school. APU tests on eleven- and fifteen-year-olds have shown that, on a number of different topics, boys perform significantly better than girls at both ages. Lynn Joffe and Derek Foxman (1986) in *Girls into maths can go* outline these results and claim that 'the main differences in performance are established by age 11'.

So the problem is not just related to external examinations. Class tests and other forms of assessment are likely to be similar in this respect. For example, if you teach in a coeducational school you may be able to compare the maths assessment results of your girls and boys. Barbara Binns (1982) did this for her third-year tutor group. Her article in *Girls into maths can go* starts with the comment: 'The results of the third-year assessment test frightened me. With few exceptions the top half were boys and the bottom half were girls.' Alan Eales (1986), also in *Girls into maths can go*, describes a similar study of the maths results in class tests, O-level and A-level in his school; he found that in *most* cases boys did significantly better than girls.

This chapter considers why girls do not seem to do themselves justice in maths examinations, tests and other forms of assessment. Before you start, you might find it useful to find out some of the views held by your pupils.

Activity 5.1

Prepare enough copies of the questionnaire in figure 5.1 (adapted from questions prepared by a teacher who helped in developing this pack) for the pupils in one of your mathematics classes – perhaps for a class about to take a public examination.

(i) Ask the pupils to complete as much of the questionnaire as is relevant to them.
(ii) Read through the completed questionnaires. For each sex

    (a) record the most common responses to questions 1 and 3;
    (b) for question 2, what (on average) is the most popular form of assessment and what is the least popular?

Figure 5.1   Maths assessment questionnaire

1. (a) Tick the statement which you feel best describes how girls and boys do in maths assessment.

☐ In general, girls do better than boys.

☐ In general, boys do better than girls.

☐ There is no difference in general.

   (b) Write down a sentence to explain why you ticked the box you did.

. . . .

. . . .

2. Put the following types of assessment in order of preference. Put a 1 in the box next to the type of assessment you most prefer, a 2 in the box corresponding to your next preference and so on. Leave a blank against these types of assessment that are not used in your class.

☐ Multiple-choice test

☐ Oral test

☐ Investigations or open-ended questions

☐ Assessment based on course-work

☐ Lots of short written questions

☐ A few longer written questions

☐ Practical work assessment

3. (a) Tick the types of question that you most like to answer; put a cross in the box next to the types of questions that you least like and leave the rest blank.

☐ Questions not involving people

☐ Questions about women or girls

☐ Questions about men or boys

☐ Questions about people in general

☐ Questions about topics which interest me

   (b) Which sorts of questions do you think you do best in? . . . .

. . . .

4. Please tick   Girl   Boy

☐   ☐

Do any distinct patterns emerge in your results, overall or with particular reference to girls or boys? Are there any surprises for you? You might like to discuss the results with your pupils.

(iii) Jot down as many other reasons as you can why you think girls may not do themselves justice in mathematics assessment.

(*Note.* The analysis of the data collected in this activity could provide an interesting statistical project for your pupils.)

Comment

When this activity was carried out in one all-girls school, most pupils felt that girls did not do as well as boys. Their reasons included:

- boys are expected to do better than girls;
- boys try harder because they need maths for their jobs;
- boys are more interested in maths than girls – girls find it boring;
- girls are slower, and need more time;
- girls get more nervous;
- girls think maths is a 'boys' subject' so they don't try hard;
- boys do more science and woodwork where they use maths;
- girls can't think so well in maths exams.

Many girls preferred course-based assessment; they disliked male-based questions, but liked those involving people in general. They also liked questions about 'topics which interest me'. How did your pupils compare? Did you notice any differences in the responses of pupils from different cultural and class backgrounds?

What about your own views? You may well agree with the views of your pupils, and you may like to consider some of the following suggestions put forward by teachers at workshops. Of course, you may have other points and you may not agree with all those listed.

- Boys are better at spotting what is required in a question. They treat maths tests as more of a game than girls do.
- External examinations are at the wrong times for girls – girls do better in maths a year earlier (fourth year).
- Teachers coach boys for exams more than girls.
- Questions often exhibit gender and cultural bias.
- Exams may fall at the wrong time in their monthly cycle for girls.
- Boys may genuinely be better at maths at age sixteen than girls.
- Maths questions often refer to practical situations related to 'boys' subjects', like physics and craft, design and technology (CDT).
- Girls are more thorough; they need more time.
- Girls lack confidence – they have too much of a desire for the right answer.
- Girls tend to panic more than boys.
- Social pressures – it is feminine to fail in maths.
- Girls don't do subjects which use maths (like physics or craft) and so do not get the extra practice that boys do.
- 'Girls' subjects' (e.g. biology, home economics) do not use as much maths as they could.
- Girls do better on more open-ended examinations such as those set for English.

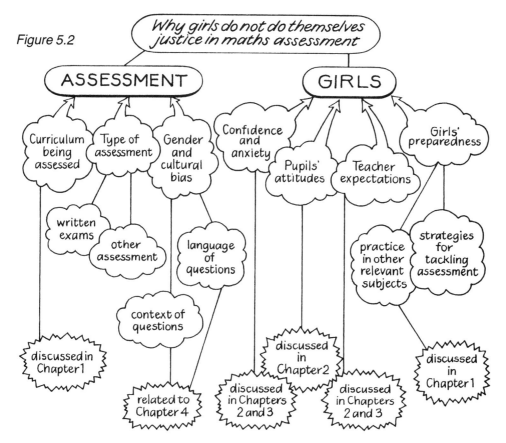

Figure 5.2

Why girls do not do themselves justice in maths assessment

ASSESSMENT

GIRLS

Curriculum being assessed

Type of assessment

Gender and cultural bias

Confidence and anxiety

Girls' preparedness

Pupils' attitudes

Teacher expectations

written exams

other assessment

language of questions

practice in other relevant subjects

strategies for tackling assessment

context of questions

discussed in Chapter 1

discussed in Chapter 2

discussed in Chapter 1

related to Chapter 4

discussed in Chapters 2 and 3

discussed in Chapters 2 and 3

Figure 5.2 shows how some of these points may be linked together. Add in your own points as appropriate.

Undoubtedly, many of these aspects have a more general influence on mathematical achievement – they are not just confined to assessment – and figure 5.2 indicates how those points have been considered earlier in pack. But the diagram indicates – as does the opening anecdote to this chapter – that the *way* in which mathematics is assessed and the *strategies* used to tackle assessment questions could also contribute to girls' lower attainment. It is this aspect of the problem that we explore here.

We begin by considering the various methods of assessment used in mathematics. In doing so we shall investigate whether different types of assessment favour different sexes. Section 5.2 then continues the theme of chapter 4 by discussing the implications for girls' performance in mathematics of including assessment questions biased in favour of one sex. In the final section we shall look at the strategies girls and boys adopt in tackling tests and consider whether girls' strategies can be improved so that they are more able to do themselves justice in examinations.

5.1 Types of assessment

Traditionally, mathematics has almost always been assessed by timed, written tests and examinations, but recently there has been much more debate about alternative forms of assessment – profiles, graded assessment and assessment by course work, project work and oral work. Could it be the case that the *type* of assessment influences pupils' achievement? For example, some kinds of assessment may be culturally biased and may unfairly disadvantage some pupils, or some types of assessment may be biased towards one sex.

One of the early pieces of work on whether the type of test disadvantages one sex was carried out by Roger Murphy (1978) on O-level examination papers, which were of two types: multiple-choice (or objective) tests and the more traditional question papers. He looked at girls' and boys' performance in the Associated Board O-level papers for a range of different subjects in June 1976 and 1977. The performance of 1000 girls and 1000 boys on the multiple-choice test was compared with their performance on the other papers for the examination. His results indicated that multiple-choice (or objective) tests seem to favour boys relative to girls. He concluded:

> This consistent male advantage in objective test papers across the whole range of subjects and across the examinations in these subjects for two consecutive years adds considerable weight to the idea that males are generally at an advantage when educational levels are measured by objective tests, rather than by other forms of assessment.

Such tests therefore tend 'to magnify an already existing male advantage' in O-level subjects like chemistry, physics and mathematics, and in some other subjects the difference is even more marked; for instance, in English language, boys' scores had a higher mean in the multiple-choice test but in the other paper girls' scores had a higher mean. Figure 5.3 summarises Murphy's 1976 results for mathematics and English language and illustrates the male advantage he discussed.

Figure 5.3   Two of Murphy's results, 1976

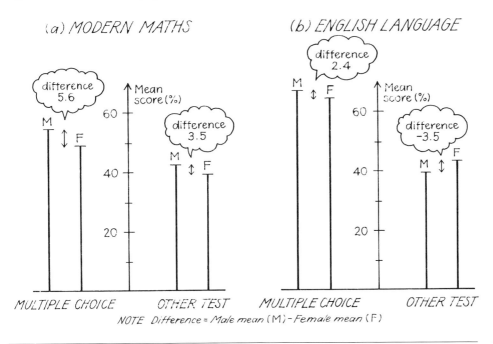

Activity 5.2

This activity invites you to collect data from one of your classes to compare the differences in performance between girls and boys on a multiple-choice test and on a comparable traditional test.

105

(i) Figures 5.4 and 5.5 provide two tests containing very similar questions. Give your pupils the multiple-choice test one week and the more traditional test the following week (or vice versa).
(ii) Calculate the mean test scores of the girls and boys on the two tests.
(iii) Compare the relative positions of the means by using a diagram like figure 5.3.
(iv) Discuss with colleagues the results you obtain in order to investigate whether girls in your school are disadvantaged by multiple-choice tests in mathematics. (Of course, one set of results does not tell the whole story but it could lead to an interesting discussion.)

Notes

(a) If you teach in a single-sex school you can still compare the performance of the girls or boys on the different types of tests.
(b) You may need to adapt the tests to suit the age-group/ability of your pupils.
(c) If your school uses both multiple-choice tests and traditional tests you may be able to obtain suitable data from existing school records.
(d) You might like to ask colleagues who teach other subjects to carry out a similar exercise. You can then compare the results for a range of subjects.
(e) You may also like to use the tests in order to investigate any cultural bias which might disadvantage certain pupils.
(f) You may like to do activity 5.7 soon after this activity.

Comment

Roger Murphy's sample was large enough for him to test his results statistically, but even smaller samples can highlight important points.

Several teachers have tried out this activity on mixed-sex classes with varying results. In many (but not all) cases girls performed worse on the multiple-choice questions compared with boys. Were your results similar?

Murphy suggested several reasons why multiple-choice (objective) tests might give males an advantage, namely:

■ males and females tend to think differently;
■ objective tests do not test verbal ability and girls have generally higher verbal ability than boys;
■ girls do not like objective tests   they may perceive them as a masculine type of activity.

Do you agree with Murphy's points?

Activity 5.3

Jot down two or three other reasons why you think girls may not do themselves justice in multiple-choice tests as a whole.

Comment

Suggestions from teachers at workshops included the following:
■ boys accept the rules of multiple-choice tests more than girls (see Brown 1984 in *Girls into maths can go*);
■ girls may get defocused, looking at a problem in a wider context;
■ boys are better at homing in on the solution quickly by eliminating unlikely answers;
■ boys are not afraid of making mistakes;

Figure 5.4 Multiple-choice test

Circle the option which gives the correct answer.

1. $1\frac{1}{8} - \frac{3}{4} =$ (a) $1\frac{7}{8}$ (b) $1\frac{1}{2}$ (c) $1\frac{1}{6}$ (d) $\frac{3}{8}$

2. $1\frac{1}{8} \div \frac{3}{4} =$ (a) $\frac{2}{3}$ (b) $1\frac{1}{2}$ (c) $1\frac{1}{6}$ (d) $\frac{3}{8}$

3. Six articles are bought; each of them costs 35p. How much change would you get from £10?

    (a) £8.10   (b) £8.90   (c) £7.90   (d) £2.10

4. $12\frac{1}{2}\%$ as a fraction in its lowest terms is

    (a) $\frac{25}{200}$ (b) $\frac{25}{2}$ (c) $\frac{1}{4}$ (d) $\frac{1}{8}$

5. What is the smallest prime number more than ninety?

    (a) 91   (b) 93   (c) 95   (d) 97

6. What is the median of 3, 4, 6, 8, 3?

    (a) 3   (b) 4   (c) 5   (d) 6

7. What is the mode of 3, 4, 6, 8, 3?

    (a) 3   (b) 4   (c) 5   (d) 6

8. A triangle has two equal sides of 3 cm and at least one angle of 70°. How many different triangles can be drawn satisfying these conditions?

    (a) 0   (b) 1   (c) 2   (d) 3

9. A ship is sailing on course 037° and turns 90° to the left. The new course is

    (a) −053°   (b) 127°   (c) 217°   (d) 307°

10. If $2x - 1 = y$ then $x =$

    (a) $\frac{1}{2}(y - 1)$ (b) $\frac{1}{2}(1 - y)$ (c) $\frac{1}{2}(1 + y)$ (d) $y - 1$

11. Please tick   Girl   Boy

        □   □

Figure 5.5 Traditional-style test

1. $1\frac{1}{4} - \frac{3}{8} =$

2. $1\frac{1}{4} \div \frac{3}{8} =$

3. Six articles are bought; each of them costs 45p. How much change would you get from £10?

4. Express $2\frac{1}{2}\%$ as a fraction in its lowest terms.

5. What is the largest prime number less than ninety?

6. What is the median of 5, 6, 8, 10, 5?

7. What is the mode of 5, 6, 8, 10, 5?

8. A triangle has at least one angle of 75° and two equal sides of length 5 cm. How many different triangles can be drawn satisfying these conditions?

9. A ship is sailing on course 046° and turns 90° to the left. What is the new course?

10. If $2x - 4 = y$, then $x =$

11. Please tick   Girl   Boy

                                     ☐   ☐

- girls are not as as willing to make an informed guess at the answer as boys;
- girls need to be sure they are right;
- boys see objective tests more as a game and they have more practice at playing these kinds of games than girls do;
- objective tests are mostly a test of speed and girls tend to be slower, especially if questions are set in a male context.

This discussion suggests that girls are at more of a disadvantage in multiple-choice tests than in the more traditional open-ended tests. But the points raised during the discussion about the reasons for this suggest that girls are actually disadvantaged by *any* sort of assessment carried out under 'exam conditions'. For example, they tend to be slower than boys, they need to be sure that they are right (see the opening anecdote) and they are more susceptible to feelings of panic and anxiety as a result of test pressure and tension.

Recent research in Holland (see Kindt and de Lange 1985) seems to confirm that girls' performance in mathematics under exam conditions may not be the best indicator of their ability. As part of the assessment on a new A-level equivalent maths course pupils were required to take independent tests under exam conditions. This work was marked in order to give a stage 1 score. Pupils were then allowed to take the paper home to work on it under more relaxed conditions. The paper was then remarked, giving a stage 2 score. For the stage 1 scores the girls' results were generally lower than the boys', but the position was reversed for the stage 2 scores. This new syllabus was also compared with a more conventional syllabus; with the conventional syllabus twice as many boys as girls passed the course whereas with the new syllabus the proportions of girls and boys who passed were roughly equal.

In Britain recently there has also been much consideration given to alternative types of assessment – both formal and informal. The discussion of these in the Cockcroft report (1982) has been taken up by the new 16+ General Certificate of Secondary Education (GCSE), and some of the aims for courses leading to mathematics GCSE are likely to provoke the development of alternative forms of assessment. For example, it is stated that courses should enable pupils to:

> 2.1 develop their mathematical knowledge and oral, written and practical skills in a manner which encourages confidence;
> 2.2 read mathematics, and write and talk about the subject in a variety of ways . . .
> (GCE and CSE Boards' Joint Council 1985)

In considering how the achievement of such aims might be assessed, the national criteria for GCSE mathematics contain *principles*, rather than prescriptions, as demonstrated by the following extracts:

> . . . It is essential that syllabuses and methods for assessment for these examinations should not conflict with the provision and development of appropriate and worthwhile mathematics courses in school . . .

> . . . assess not only the performance of skills and techniques but also pupils' understanding of mathematical processes, their ability to make use of these processes in the solution of problems and their ability to reason mathematically . . .

> . . . pupils are enabled to demonstrate what they know and can do rather than what they do not know or cannot do.

> . . . examination tasks to relate, wherever appropriate, to the use of mathematics in everyday situations.

Of particular interest here perhaps are the two specific assessment objectives that are mentioned in the GCSE mathematics criteria which can be fully realised only by assessing work carried out by candidates in addition to time-limited written examinations; namely that students should

> respond orally to questions about mathematics, discuss mathematical ideas and carry out mental calculations . . .

and

> carry out practical and investigational work, and undertake extended pieces of work.

Although there is still a requirement for one end-of-course written examination (accounting for 50% to 70% of the assessment) there will also be a requirement from 1991 for a course-work component (accounting for at least 20% of the assessment). Continuous assessment in mathematics has been used for some time in some CSE modes and it is becoming increasingly important in examinations at all levels. Furthermore, under the GCSE criteria, from 1991 oral and practical work must be included in all courses in addition to written work in order to enable teachers to provide a profile for each pupil, i.e. a detailed record of what each pupil has achieved in mathematics.

Discussion of oral, practical-work assessments and course-work profiles is now taking place and many Local Authorities are beginning to draw up plans. For example, ILEA has established a Graded Assessment in Mathematics scheme (GAIM 1985) which will require assessment records in the form of a profile organised into levels. The components of the GAIM record will be practical work, problem-solving, investigations, extended pieces of work and content criteria. The Oxford Certificate of Educational Attainment (OCEA 1985) also involves personal profiles and graded assessments, and other Local Authorities and Examination Groups may well do something similar. An interesting aspect of the OCEA philosophy is that the pupils themselves should be involved in all the main areas of certification and so maths teachers working with the pilot phases are investigating with pupils how self-assessment might be introduced into their profiles. The Association for Teachers of Mathematics (ATM) also has a working group on assessment, part of whose task is to produce an in-service pack on modes of assessment, and this will include methods of self-assessment. The first version of some activities for this pack has already been produced as the document *Working notes on assessment* (ATM 1985).

What effect – if any – will these different types of assessment have on girls' mathematical performance? Will girls fare better or worse?

Activity 5.4

There is little research evidence in Britain on other forms of assessment. Jot down what you think could be the advantages and disadvantages so far as girls are concerned regarding the introduction of the following forms of assessment:

- oral work,
- practical work,
- problem-solving and investigations,
- extended pieces of work,
- self-assessment.

Comment

Some points made by teachers at workshops include:

- Girls are not as good as boys at practical work as they do not have as much experience of it in other subjects.
- Girls present work better and are more conscientious over extended pieces of work.
- The Secondary Mathematics Independent Learning Experience (SMILE) uses course work in its assessment and this benefits girls (by about half a grade).
- Girls like investigations and do better in them than boys – in fact they seem to do better on these than on other forms of assessment.
- 'I use oral work with the lower school, especially the lower ability classes, and I find that girls are better – they are more thorough and can explain how they got their answer.'
- 'When pupils write self-assessments there is often information about my own teaching and I find that very helpful. I find girls' comments more useful than boys'.'

Although there has been almost no research about the effects of other forms of testing in this country, in Holland (Kindt and de Lange 1985) and in Denmark (Danish Ministry of Education 1985) some of the methods of assessment listed here are already included in courses in some upper secondary schools. For example, in the Dutch research, the following 'not so common tests' were carried out as part of the new syllabus 'A' mathematics assessment: oral tests; oral test with preparation (reading an article); a very open essay; a take-home test (at a considerably higher mathematical level than that encountered at school); and the two-stage test mentioned earlier. As already stated, girls and boys performed equally well on this assessment whereas on the conventional syllabus 'B' assessment 'boys dominated by 2 to 1'. Hence it would appear that girls are not disadvantaged on these other forms of assessment in the same way as they are by conventional time-tested written examinations.

On the other hand, the APU statistics (Joffe and Foxman 1986) suggest that boys are superior to girls in measuring skills so this could be reflected in any practical work in the classroom, in which case this type of assessment might again favour boys more than girls.

Finally, problem-solving and investigations cover a wide range of activities and problems. Problems of relevance to girls may not appeal to boys and vice versa. The results of SMILE examinations – which involve this type of work – are currently being analysed by sex. It will be interesting to see whether this component does benefit girls.

You may also like to consider here the possible effects of these alternative methods of assessment on pupils from different cultural and class backgrounds.

What can be done?

The discussion in this section suggests that by making changes to the *way* in which mathematics is assessed, girls' attainment might well be improved.

(1) For example, try to reduce pressure and tension (which particularly affect girls) by not continually stressing the importance of class tests, by not announcing marks in public, by avoiding surprise tests, by not making tests too difficult and by diminishing the time factor occasionally.

111

(2) Try to avoid multiple-choice tests wherever possible.

(3) Given the current structure of public examinations it is not possible simply to ignore the format of the exam paper and the time restrictions, in which case it might be a good idea to help girls develop better strategies in order to tackle these conventional types of exams (see section 5.3).

(4) The discussion here suggests that girls may well do better on alternative types of assessment, so pressure could be brought to bear on Local Authorities and Examining Bodies to make changes to their assessment strategies as quickly as possible (see figure 5.9 in section 5.2).

(5) Find out more about the sort of practical work and investigations that could be carried out in the maths classroom. If possible, become involved in the development of this sort of work. Try to ensure that any practical work and investigations are equally relevant to boys and girls.

(6) Find out more about oral work and try to use it in the classroom. You might find the following ideas useful (these were put forward at workshops by teachers who are already trying out oral testing):

- 'Oral tests can test different things than written tests; you can ask a pupil to explain why they got a particular answer.'
- 'First I ask what topic they have enjoyed most and get them to talk about it, then I give them a question, tell them to read it and then talk about it before doing it.'
- 'I use oral work for revision; I ask questions like "tell me all you know about angles".'
- 'I had a talk with the French teacher who is used to oral assessments; she helped me to get children to feel comfortable and less nervous.'
- 'Using prompts and open-ended questions like "how did you do that?" are useful.'
- 'Oral tests are more time-consuming; I do them regularly but shy children suffer so they are better done "privately".'
- 'Girls are better at expressing themselves than boys in oral work and in written explanations.'

(7) Find out more about course profiles (see the further reading and additional information list at the end of the chapter). Again you might like to become involved in the development of this work.

5.2 Bias in questions

In order to meet content criteria and to use course-work in assessment, worksheets and other materials are often produced as part of assessing pupils' progress. For example, the Hertfordshire Mathematics Achievement Certificate (Herts Maths Advisory Teachers 1985), which is designed for pupils whose needs are not met by existing O-level and CSE courses, requires pupils to complete a number of work topics with associated worksheets. The sheets meet the GCSE aims of using mathematics in everyday life; they have titles like 'Till checking', 'Weight and price of paper to make envelopes', 'Orders and sales', 'Using sines and cosines in an engineering factory', 'Co-ordinate measuring machine', 'What containers does the canteen need today' and 'Canteen orders'. Many examining bodies may move over to assessment like this, which concentrates more on using maths in context. But there are safeguards which need to be made in order to ensure that some pupils are not disadvantaged in any way. If 'everyday life' means 'everyday life for a white male' then the situation for girls and also for boys from some cultural backgrounds will remain the same. In sections 4.1 and 4.2 you have already examined textbooks and teaching materials for male and female contexts and for gender and possibly cultural bias. Richard Graf and Jeanne Riddell's work (1972) was mentioned, showing that females take longer to solve problems set in a male context, and a study by Muriel Eddowes *et al.* (1980) showed that girls do not perform as well as boys on such questions. However, in chapter 4 you looked at bias in mathematics materials *in general*; now we ask you to think more specifically about *assessment* questions since children may well be affected by the context of such a question being unfamiliar or inappropriate to them.

Activity 5.5

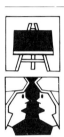

(i) Look at the maths assessment questions set in your school or taken by your pupils in public exams. Count how many are set in male, female, neutral or unfamiliar contexts.

(ii) Ask your colleagues to do the same. Discuss the balance and what you can do about it.

Comment

Feedback from teachers who have done this, and inspection of several external examination papers, suggests that 'male' contexts appear far more often than female contexts. For example, a recent London CSE paper had six male contexts but no female contexts, although most questions were context-free. One teacher found a ratio of 16 to 5 male to female contexts in Associated Board papers her pupils had just sat. She said, 'I am rather ashamed to admit that I never noticed this bias when giving these exam papers to students.' Another teacher looked at a test which he was about to give to his pupils and noticed some bias. He changed one context from a motorcycle shop to a dry cleaners to try to redress the balance. It did not seem fair as it was.

One teacher looked at a question on a lathe (on an external exam paper), which had put off girls in her group because they knew little about lathes. On closer examination of the question, she discovered that there was no need to know anything about lathes in order to do the question and that, had the girls not been put off, they could probably have done the question anyway.

With most assessment questions it is relatively easy to decide whether the context is male or female, or neutral. A naval officer, a striker for a football team or a lathe problem are clearly 'male', whereas mothers and their children, food purchased by a housewife and completing a six-month typing course are clearly 'female' contexts (although completing a word-processing course could possibly be considered 'male'!).

Even when there is no obvious male (or female) context (say) in a question like 'A man buys six articles at 35p each. The change he receives from £10 is . . .', the person mentioned is male and this may also reinforce the feeling that maths is a male subject, as demonstrated in section 4.1.

Many people feel that sexist assumptions in language affect pupils' examination performance. For example, Lesley Kant (1982) suggests that any study of science and mathematics syllabuses will show that examples tend to be related to traditional male interests and activities and that the applications to areas of possible female interest are rare. She concludes that the language of exams 'tends to assume that examinees and populations are male with constant references to "he", "him" and "man".'

During the development of this pack, one teacher decided to determine for himself the extent to which the context of the test affected the results of girls and boys. He devised two similar tests testing similar mathematical concepts. These are illustrated in figures 5.6 and 5.7. In one test the contexts and people are mainly female and in the other they are mainly male. As a control, one question on each test is the same and is set in a neutral context. The tests were given in one lesson period (40 minutes) under semi-exam conditions to mixed parallel classes and the results were then analysed.

Activity 5.6

This activity invites you to carry out your own research on whether the context of a test makes a difference to your pupils' performance. You might like to use the tests in figures 5.6 and 5.7. However, these tests may not be appropriate for your pupils for cultural reasons or because of the assumed level of mathematical ability, and you may therefore wish to devise your own 'female' and 'male' tests.
  (i) Use the tests with a group of children of comparable ability. Divide both the girls and the boys into two equal groups. Give the 'male' test to one group of boys and to one group of girls. Give the 'female' test to the other groups.
  (ii) Calculate the mean test scores for each group of pupils and compare your results. Have a look at the individual question responses. Does the context and/or the sex of the people mentioned have any effect on how well girls (and boys) perform on the test?
(*Note.* You might like to do activity 5.7 immediately after this activity.)

Comment

In the experiment mentioned earlier, the tests were marked out of twenty; the teacher's results are illustrated in figure 5.8. All the boys got the football league question right but some of the girls got it wrong. Everybody got question 2 correct on the female paper.

Figure 5.8   Scores of one class on the tests in figures 5.6 and 5.7

	Female test mean	Male test mean
Girls' scores	13	$10\frac{1}{2}$
Boys' scores	16	$15\frac{1}{2}$

How do your results compare? One teacher who did a similar exercise with fourth and fifth years found that boys complained about 'female questions' whereas girls just got on and did the male questions (why do you think this is so?). However, girls did 12% better on the female paper and completed it quicker. The boys' marks were very similar on both papers.

Of course, it is not always easy to decide whether a question is actually biased (say) against one sex. The context may be relevant and also the sexes of the people mentioned, but there may be other factors to consider; for instance, girls seem to find questions about people easier, whilst boys prefer those relating to things.

Barbara Strassberg-Rossenberg and Thomas Donlen (1985) tried to classify test items for male/female bias. They looked for questions which females or

Figure 5.6   Test paper 1

1. *LEEK MOUSSE*
    500 g of leeks
    125 g of butter
     50 g of flour
    4 eggs

    Leeks cost 80p per kilogram (1000 g), 250 g of butter costs 42p, 1 kilogram of flour costs 40p, and eggs are 84p for 12. What is the cost of ingredients for this dish?

2.   *HOW MANY TIMES*

	Cinema	Disco	Visit friends	Cost for month
Jane	4	2	1	
Joan	5	1	3	
Janet	3	4	3	
Joanna	2	2	5	

    This table shows the way 4 girls have spent some of their time in the last month. It costs £3 to go to the cinema, £1 to go to the disco, and nothing to visit friends. Copy the table and complete the 'cost for month' column.

3. Some money is shared between Mandy and Kim. Mandy gets £84 and Kim gets £14. How many times more has Mandy than Kim?

4. A shop has a sale and advertises that all its dresses have been reduced by 10% and all the shoes have been reduced by 5%. What will be the cost of the following?
    (a) a dress which was £35
    (b) a pair of shoes which were £20
    (c) a dress which was £17.56
    (d) a pair of shoes which were £24.75

5.

Tonbridge	d				08 39		08 43				09 00				09 23		09 35		09 44
Hildenborough	d						08 47				09 05				09 27				
Sevenoaks	d			08 36			08 54	08 57		09 14	09 17	09 25			09 35	09 37	09 45		09 57
Dunton Green	d			08 40			09 00				09 20					09 40			
Knockholt	d			08 46			09 06				09 26					09 46			
Chelsfield	d			08 48			09 09				09 29					09 49			
Orpington	a			08 51			09 12		09 23	09 31				09 45	09 54	09 55		09 54	10 04
Orpington	195 d		08 44	08 52		08 55	09 12		09 17	09 23	09 32			09 37	09 46	09 57	09 55		09 57	10 06
Petts Wood	195 d		08 47	08 55		08 58	09 15		09 19		09 35			09 39				09 59
Chislehurst	d		08 50			09 01			09 22				09 42			10 02
Elmstead Woods	d		08 52			09 03			09 25				09 45			10 05
Bromley North	d	08 47					09 07		09 20				09 40			10 00
Sundridge Park	d	08 49					09 09		09 22				09 42			10 02
Grove Park	d	08 52	08 55			09 07		09 12	09a25	09 28			09a45	09 48		10a05	10 08
Hither Green	201 d	08 56				09 10		09 16		09 32				09 52			10 12
Lewisham	199, 201 d																
St Johns	199, 201 d																
New Cross ⊖	199, 201 d					09 15				09 37				09 57			10 17
London Bridge ⊖	199, 201 a	09 07				09 21		09 26		09 45	09 39	09 52		10 03			10 23
London Cannon Street ⊖	199, 201 a			09 15	09 19	09 25			09 35			09 57	09 52	10 07	10 05		
London Waterloo (East) ⊖	199, 201 a	09 14	09 12				09 23	09 31		09 51	09 44				10 14	10 28	10 27
London Charing Cross ⊖	199, 201 a	09 17	09 15				09 26	09 34		09 54	09 49				10 17	10 31	10 32

    Use this timetable to answer the following questions:

    (a) What is the fastest train from Grove Park to Charing Cross?
    (b) What is the slowest train from Grove Park to Charing Cross?
    (c) What time does the 0907 from Bromley North arrive at London Bridge?
    (d) How long does this journey take?
    (e) If you get the 0940 from Bromley North, what time will you arrive at London Bridge?
    (f) How long does this journey take?
    (g) How long does it take to travel from Orpington to Grove Park by train?

6. Please tick   Girl   Boy

                      ☐   ☐

115

Figure 5.7 Test paper 2

1. A garage services a car and uses the following:

   5 litres of oil   50 cc of gear oil
   125 g of grease   4 spark plugs

   Oil costs £8 for 10 litres, grease costs £2.50 for a 250 g tin, gear oil costs £4 per litre (1000 cc), 12 plugs cost £6.60. What is the cost of materials?

2.

	Won	Drawn	Lost	Points
Millwall	4	2	1	
Arsenal	5	1	3	
Tottenham	3	4	3	
Wimbledon	2	2	5	

   This is part of a football league table. A team gets 3 points for a win, 1 point for a draw, 0 points for losing. Copy the table and complete the points column.

3.

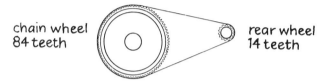

   chain wheel 84 teeth   rear wheel 14 teeth

   This diagram shows how the chain wheel and back wheel on a bicycle are arranged. How many times will the back wheel go round for each turn of the pedals?

4. A sports shop has a sale and advertises 10% off all clothes and 5% off all equipment. What will be the cost of the following?
   (a) football boots which were £25   (c) football shorts which were £7.56
   (b) a cricket bat which was £20   (d) a football which was £17.75

5.

Tonbridge	d					08 39		08 43				09 00				09 23		09 35			09 44
Hildenborough	d							08 47				09 05				09 27					
Sevenoaks	d				08 36		08 54		08 57			09 14	09 17	09 25		09 35		09 37	09 45		09 57
Dunton Green	d				08 40		09 00					09 20						09 40			
Knockholt	d				08 46		09 06					09 26						09 46			
Chelsfield	d				08 48		09 09					09 29						09 49			
Orpington	a				08 51		09 12			09 23	09 31				09 45		09 54	09 55		09 54	10 04
	195 d		08 44		08 52	08 55		09 12		09 17	09 23	09 32			09 37	09 46		09 57	09 55		09 57	10 06
Petts Wood	195 d		08 47		08 55	08 58		09 15		09 19		09 35			09 39					09 59
Chislehurst	d		08 50			09 01				09 22					09 42					10 02
Elmstead Woods	d		08 52			09 03				09 25					09 45					10 05
Bromley North	d	08 47					09 07		09 20				09 40					10 00		
Sundridge Park	d	08 49					09 09		09 22				09 42					10 02		
Grove Park	d	08 52	08 55			09 07		09 12	09a25	09 28			09a45	09 48			10a05	10 08
Hither Green	201 d	08 56				09 10		09 16		09 32				09 52				10 12
Lewisham	199, 201 d																	
St Johns	199, 201 d																	
New Cross ⊖	199, 201 d					09 15				09 37				09 57				10 17
London Bridge ⊖	199, 201 a	09 07				09 21		09 26		09 45	09 39	09 52		10 03				10 23
London Cannon Street ⊖	199, 201 a			09 15	09 19	09 25		09 35				09 57	09 52		10 07	10 05
London Waterloo (East) ⊖	199, 201 a	09 14	09 12				09 23	09 31			09 51	09 44			10 14		10 28	10 27
London Charing Cross ⊖	199, 201 a	09 17	09 15				09 26	09 34			09 54	09 49			10 17		10 31	10 32

   Use this timetable to answer the following questions.
   (a) What is the fastest train from Grove Park to Charing Cross?
   (b) What is the slowest train from Grove Park to Charing Cross?
   (c) What time does the 0907 from Bromley North arrive at London Bridge?
   (d) How long does this journey take?
   (e) If you get the 0940 from Bromley North, what time will you arrive at London Bridge?
   (f) How long does this journey take?
   (g) How long does it take to travel from Orpington to Grove Park by train?

6. Please tick   Girl   Boy

   ☐ ☐

males found significantly harder. Their results are interesting. Males did better on questions set in contexts such as science, traditional male interests or skills, transport and communications, the relationship between space and time, political science, electricity. Female contexts included cooking, clothing, personality characteristics. These categories are hardly surprising. So, since there appear to be far more contexts which favour 'males' in maths assessment questions in general, perhaps it is not surprising that males perform better. And, as mentioned in section 1.1, one of the reasons that girls doing physical science or technical courses do better in maths O-level could be that they have more background knowledge to help them answer 'masculine' questions. You may recall that Shiam Sharma and Roland Meighan (1980) suggested that pupils doing traditionally maths-related subjects have more practice at questions set in a male context and so are in a better position to tackle maths questions set in male contexts.

What can be done?

(1) Chapter 4 contains a number of suggestions aimed at reducing the bias in teaching materials as a whole, and all of these apply also when devising examinations, tests and other forms of assessment.

(2) As in chapter 4, all these suggestions can quite easily be put into practice when devising your own assessment materials, but what can be done to prevent gender or cultural bias in public examinations? After working through a draft of this chapter, one group of London teachers produced the checklist of questions shown in figure 5.9. They sent this list, together with a covering letter, to the London Regional Examining Board who circulated the list to all members of its moderating committee. You may like to do something similar – even if you do not agree with the particular points raised by the London teachers. Discuss with your colleagues what you mean by a question or test being gender- or culture-biased and make your own checklist of points. Even if this is not used to exert pressure on Examining Bodies and Local Authorities it could form the basis for a school policy on assessment.

(3) Another possibility is to broaden girls' experience so that they can cope better with 'male' contexts. One teacher tried this by using a test question on petrol consumption (which had thrown some girls) in order to discuss how a car worked in the hope that such contexts would be more familiar to them in future (see also section 5.3).

5.3 Strategies in tackling assessment

While changes are taking place, girls still have to cope with assessments which may disadvantage them; hence they need to be prepared to tackle them. Changing the assessment is only one half of the story. Improving girls' strategies in *tackling* assessments is the other. For there is evidence that girls may not have such good examination strategies overall as boys. For example, Muriel Eddowes *et al.* (1980) suggested that girls tend to try to recognise the situation and then 'apply the rules to recognised situations' whereas boys tend to answer questions 'using more independent processes – often successfully – in the more difficult questions'. They also discovered that 'those girls who did identify the problem were more reliable in calculation than the corresponding boys'. Girls more than boys are likely to be thrown by an unfamiliar context.

Figure 5.9 Questions about examinations (*Note.* These points also refer to materials other than assessment questions, e.g. textbooks, worksheets (see chapter 4).)

(i) Style of examination

Question	Reason
■ Is it multiple-choice or are there multiple-choice questions at the beginning?	Girls often do less well on multiple-choice tests.
■ Is it tightly timed?	Girls are more likely to feel anxious because of pressure of time; they may take longer to solve problems set in a male context.
■ Does it take course-work into consideration?	Evidence is emerging that girls are doing better on course-work assessments.
■ Does it include oral questions?	Girls seem to do better in oral work.
■ Does it appear 'friendly' in design?	Girls are often more nervous.
■ What is the order of difficulty of questions?	Are they ordered in terms of difficulty to give girls (and boys) more confidence?
■ Is the mathematics in familiar contexts?	Girls take longer to solve problems in unfamiliar contexts.

(ii) Gender bias in questions: points to look for

- ■ Are there more references in the text or in illustrations to males than females?
- ■ Do neuter references imply male context?
- ■ Are the females referred to by their first names whilst males are given titles, (say) Mr Brown?
- ■ Do the references to people portray them in stereotyped roles such as workmen, women shopping etc.?
- ■ Are females in passive roles and males in active ones?
- ■ Are there questions involving physics formulas or CDT which are likely to disadvantage girls because they are less likely to be familiar with them?
- ■ Are questions set in contexts unfamiliar to girls, for example at cricket matches, at factories?
- ■ Does the language of the questions clarify or confuse?
- ■ Is the vocabulary male-oriented?
- ■ Do diagrams refer to things which are closer to the experience of boys, such as a cross-section of metal girder?
- ■ Is there a preponderance of questions involving special concepts with which girls have less experience?

Another aspect, researched by Robert Wood (1976) in a 1973 O-level maths exam, showed that girls were more likely to 'choose answers which are far too large or small' and that they were more likely to miss the final stage in a problem of converting their mathematical answers into answers to the question set. For example, in a pie chart question, they might find the missing angle but not convert it into an equivalent number of children represented by it. This suggests that girls are less likely to look at their answer to a question and say 'Is this a sensible answer?'. This could be related to their lack of confidence, or it could be due to the way that they learn and are taught.

Activity 5.7

This activity may be most suitable around the time pupils are taking exams or just after a test (as in activity 5.2 or 5.6).

(i) Ask your pupils to write down how they go about tackling assessment questions of different types. Ask them what is the first thing they do when faced with a question and see what their strategies are.
(ii) Next, ask them about their *overall* strategies for tackling tests and exams.
(iii) Discuss with pupils your own strategy for tackling exams and, if convenient, discuss with colleagues what theirs are.

Comment

Several teachers asked their pupils about their strategies for tackling the type of test questions in activities 5.2 and 5.6. Most pupils stressed simply 'reading the question'. As for *overall* strategies, all pupils tended to answer the questions in the order in which they came on the paper and just plod through them. However, more boys than girls seemed happy to give up on a question when they got stuck, leave it and go to the next one, returning to it if they had time. More boys than girls eliminated 'wrong' options from multiple-choice questions and made an informed guess if they were unsure. More girls than boys mentioned trying to make sense of the question. One girl wrote 'I looked at question 8 and couldn't understand any of it. I talked to myself about the question and found that I could make sense of it.' Hardly any pupils mentioned checking or drawing helpful diagrams; the few that did were boys.

Overall, most teachers felt that few of their pupils' strategies were particularly good and that the girls' strategies tended to be worse than the boys. One teacher commented that 'exam technique' was not usually taught except in the context of external exams. Is this the case in your school?

There are many theories as to why girls and boys might approach problems differently. For example, you may recall the discussion in section 3.3 about 'serialist' and 'holist' learning strategies (see Scott-Hodgetts' article in *Girls into maths can go*). We mentioned there that serialists – who are more likely to be female than male – are often anxious about unfamiliar situations. They therefore tend to be better at assessment of familiar concepts in familiar contexts. On the other hand, holists – who are more likely to be male – may do better on unfamiliar questions.

Another theory is outlined by Stephen Brown (1984) in *Girls into maths can go*. He suggests that there are two systems of thought: one more appropriate to the humanities, the other to the sciences. Females tend to adopt the former approach to problem-solving and males the latter. Females, he suggests, tend to be more responsible and caring and concerned with 'correctness among people'. They are more likely to request more information or to locate the episode in a broader context and to reject the original question. Many males on the other hand accept the situation as posed and the 'taken-for-granted reality' implicit in it, and come up with an answer without querying the context or the question. This is likely to make quite a difference to the speed at which questions are answered. So in a timed-exam situation the 'female' approach is likely to be at a disadvantage. He also suggests that while males are happy to look for general principles to be used in future cases, females are disinclined to do this – preferring to treat each context independently. So again the 'male system' tends to be quicker in unfamiliar situations, because the general rules can be applied regardless of the context.

Both these theories suggest that typical 'female' approaches to learning and problem-solving disadvantage girls in maths assessment. Consequently, it is argued that the overall approach to learning and the types of activities encouraged in the classroom are inextricably linked to successful strategies for tackling assessment. The discussion in section 3.3 is therefore particularly relevant here.

What can be done?

(1) Many of the suggestions put forward in section 3.3 are relevant to general classroom activities including assessment, and you may like to look back at these now.

(2) Other points specifically raised in this section relate to exam technique and the strategies used to tackle individual questions. You might therefore like to make a list of good strategies for different tests (including multiple-choice and longer questions). Some strategies suggested by teachers and colleagues are given in figure 5.10. You might like to add your own ideas to this list.

Figure 5.10   Some useful strategies for tackling exams

- Practise exam-type questions as much as possible as part of your revision; you get better (and faster) by doing, not by reading.

- Try to time yourself and keep roughly to a predetermined timetable in an exam, based on the number of marks per question.

- Read through the paper quickly first and put a tick by the questions you think you can do most easily. Start with these.

- Do the questions you know you can do first; leave ones you might have trouble with until last.

- Don't get hung up on one question. If you get stuck, leave it and come back to it later if you have time.

- If you can't do one part of a long question you may still be able to do later parts.

- Don't give up if you can't understand a question. Leave it and come back to it later; it may make sense then.

- If you can't visualise what is meant by a question, a diagram may help.

- Eliminate options which are obviously wrong in multiple-choice questions; and if you don't know the answer, make an intelligent guess from the remaining ones.

- Try to estimate the rough size of the answer so you know if you are way out.

- Look at your answer. Ask 'Is this sensible, and does it answer the question?'

- Try to build in checking as part of your way of working to make sure you are on the the right lines.

- Good problem-solving strategies (for example, estimation and approximation for checking an answer) are also good exam strategies.

5.4   Summary and reflection

The problems discussed in this chapter were divided into two areas. The first concerned the *form* that assessment takes, and we suggested that it may be possible to change this so that no pupils are disadvantaged either by the *type* of assessment or by any *bias* in questions. The second area concerned the more general attitudes of the girls themselves and in particular their strategies for preparing for and tackling assessment questions. We suggested that improving girls' attitudes and expectations and helping them to develop better learning strategies and exam technique could also help girls to do themselves justice in assessment.

Two last points. First, assessment is one area in which things change slowly and where it is difficult to be effective on your own, so it might be a good idea to involve all the members of your department in trying to create a whole-school policy. Second, although assessment influences what is taught and how it is taught, the examination room is not the place to startle people for the first time with thought-provoking role-reversals. Discussion of assessment inevitably leads to looking at the curriculum and what is being assessed. It is therefore important to tackle bias in assessment as part of an *overall* policy for equality throughout the teaching of mathematics.

Activity 5.8

Jot down in the table below the aspects of the problems discussed in this chapter which are most relevant to your pupils or to your school. Beside each of these, note the strategies you feel you can most effectively introduce in order to improve the situation either now or in the future.

Aspect of the problem	Strategies for tackling it

Further reading and additional information

- Articles by B. Binns, A. Eales, R. Scott-Hodgetts and S. Brown in *Girls into maths can go*
- For further information on alternative methods of assessment you might like to consult the following publications:
  - GAIM, *Newsletters*, available from The GAIM Office, 552 King's Road, London SW10 0UA
  - OCEA, *Newsletters*, available from Research Department (OCEA), University of Oxford, Delegacy of Local Examinations, Ewert Place, Summertown, Oxford OX2 7BZ
  - ATM, *Working Notes on Assessment*, available from ATM, Kings Chambers, Queen Street, Derby DE1 3DA

Where to go from here?

(1) We hope that by studying this pack you have been able to identify and explore those aspects of the problems faced by girls in connection with mathematics that are particularly pertinent to your school and to your own teaching situation. We hope too that the suggestions listed in the 'what can be done?' sections provide useful strategies for you to employ when trying to resolve these problems. Inevitably, though, it takes time to bring about changes in attitude and in teaching-practice, so it is likely that much of this work will be ongoing.

(2) You might find it useful to keep in touch with other people who are interested in gender issues. One way of doing this is to join the Gender and Mathematics Association (GAMMA). Details can be obtained from Marion Kimberley at the Department of Mathematical Science, Goldsmiths College, New Cross, London SE14 6NW. Other possibilities include the Women in Education Group (WEG) and discussion groups associated with the Girls into Science and Technology project (GIST). Alternatively, there may be a local association or discussion group that you could join (or you might like to set up such a group within your school or area, involving both maths and other subject teachers).

(3) You might like to run a 'Be a Sumbody' conference for girls in your school or local area. These one-day conferences have been held for twelve- to fourteen-year-old girls in order to provide a different view of mathematics from that usually encountered at school, and to expand awareness of the role of mathematics as a critical filter into future employment. Each day is structured around two maths workshops, in which activities are deliberately group-centred, creative and challenging, and a careers fair in which the girls meet women in non-traditional jobs. A report on these conferences can be found in Leone Burton and Ruth Townsend's article 'Girl Friendly Mathematics' in *Girls into maths can go*.

(4) Throughout the pack we referred to the research that has been carried out on gender issues, but these references have been necessarily brief. You may therefore wish to follow up the studies mentioned by reading the articles and books in the bibliography.

(5) Many of the reasons for girls' under-achievement in mathematics are related to the ways in which mathematics is taught and assessed in school. Individual teachers can make changes to their own practice but this in itself is unlikely to be sufficient. Pressure could therefore be exerted on schools and Local Authorities to support whole-school policies in this area. You might also like to lobby local and national maths education groups in order to encourage them to reassess their ideas on teaching and assessment.

(6) The activities in this pack may have suggested ways in which you might explore other equality issues. As mentioned in the foreword, there has been very little research on the influence of race and social class on mathematical attainment. You may now feel that you can commence your own small-scale research projects on such influences and these might – in the future – provide the basis for other packs such as this.

Appendix
The computer and mathematics

As indicated in the introduction, this appendix is included because of the many connections between mathematics and computing. For example, many teachers of mathematics also have responsibilities for computing, school computers are frequently sited in the maths department, and computers are often used more in maths classes than in other lessons. The appendix also highlights the fact that mathematics is not the only subject area where girls tend to fare worse than boys. Girls face similar problems in other traditional 'male' subjects (see, for example, figure 1.4 for an indication of girls' performance in other subjects), and unfortunately the relatively new field of computing – in all its forms – is already beginning to fall into this category. Here we briefly investigate the problems faced by girls in computing and in using computers. We conclude that while it may be desirable to sever some of the links between the two areas, strengthening others could have a positive effect on girls' attitudes towards both computers *and* mathematics.

A.1   What is the problem?

> I did computer studies because I thought it would help me get a job when I leave school. But it's all so boring and nothing seems to be relevant to the real world.
>
> (Debbie, aged 15)

> I'm no good at maths, so I doubt if I'd be any good with computers. That's one reason why I don't go to the computing club at school . . . Another reason is because none of my friends go I suppose . . .   (Jasmine, aged 14)

> I don't often get to go on the computer . . . None of the girls in my class gets much chance because the boys are always crowding round it all the time . . . Anyway, I don't really like it that much when I *do* get a turn . . .   (Lynette, aged 11)

These anecdotes suggest that girls are already beginning to develop negative attitudes towards computers and computing in spite of the relatively recent introduction of the 'new technology' into many schools – attitudes which are similar to those that are demonstrated towards mathematics.

Of course, some people do not regard the reluctance of girls to become involved with computers as a problem; for instance, consider the article in figure A.1 on page 124. But this perhaps illustrates a rather stereotyped point of view and it could be argued that Sarah Goodwin presents a false dichotomy, for is there really such a sharp male/female division between on the one hand logic and reasoning and on the other the ability to generalise and to develop theories? The article also fails to consider any economic factors which are surely of considerable importance!

Figure A.1 (Goodwin 1984)

Women's intuition

SARAH GOODWIN

Our all-girls school has recently acquired a computer network. It is under-used, and most staff and girls would not notice if it disappeared.

Female apathy towards computers, which seems to be in the same tradition as their distaste for science and technology, provokes deep concern among some educationists, expressed in renewed efforts to induce girls to overcome early socialization or "conditioning" and become more like boys. I believe that the concern is misplaced and will in the end do great damage to the education of both boys and girls.

Education is about developing intelligence, if we take that word to mean the ability to understand and operate within our environment. There are two sides to this intelligence, both equally important.

One side is mechanical. It includes the skill of reasoning, using the rules of logic. It can develop most easily within artificial formal systems like mathematics, games like chess and computer programming. The material world can be understood as following formal laws, so there seems to be a direct reward to be gained from using this side of intelligence to predict and control events. In a culture that links masculinity to power and action it is small wonder that boys take to subjects which allow them to identify with the image.

The other side to intelligence is evaluative and creative. It is the side which generalizes from the patterns of reasoning, to develop theories, imagine outcomes and create alternatives. It uses intuition and "feel". This side is, of course, involved to some degree in the manipulation of formal systems to the extent that they are understood rather than just worked within.

It is, however, developed most fully in areas relating to the social environment, where the mechanical patterns are already "given" and do not have to be constructed (the rules of language, for instance). The social sciences, humanities, art and music all require and develop this side of intelligence, and the reflective, passive way in which study of these subjects is carried out leads them to be associated with low-status effeminacy.

No prospect of control exists, only affective reward, so boys enter this area of study at the risk of ridicule, and come to it late if at all when the conflict with masculinity can be resolved without threat to their self-image.

Girls are being persuaded to broaden the development of their intelligence by embracing the "masculine" side of logic and reasoning. While I would champion their right to choose this course, I cannot but notice the absence of any parallel attempt to persuade boys to strengthen their intelligence by a complementary crossing of the boundary. If girls can only be recognized as intelligent by developing that side of intelligence normally over-developed in boys, they may desert and compete in a man's world. Perhaps they will beat the men at their own game, but it is more likely that they will end up playing second fiddle.

Girls can excel in the sphere of the traditional side of their intelligence, and they should not be induced to exchange their birthright for a mess of pottage. They may, even now, be the only ones with the wisdom to understand this as they vote with their feet to stay out of the computer room. One day men too will understand that the social environment is at least as important and infinitely more fascinating to study. Perhaps we should start to persuade them now, before it is too late.

Sarah Goodwin is head of social studies at Ravensbourne School for Girls, Bromley.

Activity A.1

Jot down one or two reasons why you think girls should be encouraged to become involved with computers.

Comment

Perhaps the most obvious reason is simply that computers play a fundamental role in this age of technology and if girls do not become involved they will be left behind. Computers are tools which *all* pupils should be able to use in many contexts. More and more jobs depend on them in manufacturing, office work, banking, the Health Service and so on, and there are already signs that women are losing out. For example, a study funded by the Equal Opportunities Commission (EOC) has established that more women than men are losing jobs through high technology.

> The survey, conducted among 40 West Yorkshire companies, showed that technology – automation, computer control, electronics and the like – is having an adverse effect on the kind of work in which a large proportion of women are employed. The study also showed that the new jobs which were being created were in scientific and technical areas, where fewer women than men were qualified.

(Johnstone 1984)

Also, it is likely that school-leavers who are confident in using computers will be better prepared to venture into jobs which involve other sophisticated machinery as well as computers, again suggesting that it is important for all pupils to become involved with computers at school.

Teachers at workshops mentioned non-economic factors as well. For example, they argued that school computing needs a better image and that more female involvement could have a very beneficial effect. They also questioned the effect of non-involvement on girls' self-images and confidence; it was stressed that non-involvement in any important area of the school curriculum must have a negative effect.

The above discussion suggests that *all* pupils should participate equally in *all* types of computer-related activities, but evidence suggests that there is still a long way to go before this goal can be achieved.

Activity A.2

What is the position in your school?
(i) Collect data from your tutorial group or from one of your classes to complete the table in figure A.2. (You might also like to ask a colleague to complete a similar table for a different age-group.)
(ii) Look at the percentage columns. What can you conclude? Are your conclusions the same for each age-group or can you identify any changes as children get older?

Figure A.2 Pupils' participation in computer-related activities

	Number of girls	Number of boys	Proportion of all girls in group	Proportion of all boys in group
At school 1. Pupils using computers in (a) maths lessons				
(b) other lessons (which ones?)				
*2. Pupils taking courses in: (a) A-level computer science				
(b) O-level/CSE computer studies				
(c) pre-vocational course				
(d) other (e.g. computer-appreciation)				
3. Pupils in computing club				
Outside school 4. (a) Pupils with access to a home computer				
(b) Pupils with exclusive use of a home computer				
5. Pupils who regularly read a computer comic or magazine				

 * Complete as much of this part of the table as appropriate.

Comment

We hope that you did not find any differences in the proportions of girls and boys using the computer in maths lessons (although we shall return to this point later), but we suspect that you observed higher proportions of boys than girls in all categories where participation is not compulsory – and across the age-range. Furthermore, the differences are likely to increase with age. So once again (as with mathematics) it seems that girls are 'opting out' when they have the opportunity. The situation seems to be worst in mixed secondary schools although you may also find evidence of similar trends in junior schools or in single-sex schools. Compare your findings with the following information.

Courses in computing

DES statistics in figure A.3 show that fewer girls than boys are entering public examinations in computing. This pattern of subject take-up is partly as a result of option schemes which automatically discriminate against girls wanting to take computer studies alongside (say) French or German, but it is also because computer studies – like physics or chemistry – tends to be considered as a maths-related subject. In fact, computer studies has a particularly poor image and this is picked up later in the chapter.

Computer club

There is considerable evidence that fewer girls than boys are participating in extra-curricular computing activities. For instance, a survey of the 23 schools with fourteen- and fifteen-year-olds in Croydon estimated that only 6.6% of those attending computer clubs at lunchtime and/or after school were girls, and that the proportion of girls decreased as they became older (see EOC 1983).

Figure A.3 Public examination results in computing (England only; summer 1983; adapted from DES 1984)

Examination	Number of entries		% of entry gaining pass grade/graded result*		% of entry gaining highest pass grade	
	Boys	Girls	Boys	Girls	Boys	Girls
CSE computer studies	23 885	15 444	89.47	91.02	12.06	8.77
O-level computer studies	30 527	13 222	60.60	51.34	8.69	5.11
A-level computer science	5 163	1 409	69.79	62.38	4.67	1.85

* CSE: graded result (grades 1–5). O-level: pass grade (grades A–C). A-level: pass grade (grades A–E).

Access to home computer

Girls are also less likely to have a computer in the home. The same survey found that in one school of 1200 pupils with a very mixed catchment area, 109 had access to a computer at home and that only 30 of these were girls, and of the 64 pupils with exclusive use of a home computer only 13 were girls. Robin Ward (1986) in *Girls into maths can go* reports that at least *nine times* as many boys as girls are likely to have a computer in their homes, and an investigation by the Acorn Computing Company of Great Britain (Grady 1983) obtained the even more extreme ratio of 13:1. This imbalance is not confined to secondary-school children. As might be expected, girls who have brothers seem to be better off than those who do not. For example, Anita Straker (1986) in *Girls into maths can go* concludes that at least in primary schools this problem is most acute for girls from girls-only families (see activity 4.8). There could also be particular problems for girls from some cultural or economic backgrounds. Can you identify any of these problems in your school?

It is not altogether clear, however, that girls do not want home computers; for instance, consider the following anecdote:

> The problem was brought home to me by the experience of children in a class of nine-to ten-year-olds that I was teaching. All but one of the 26 children in the class said before Christmas that they would like a micro. Most of the children in the school came from fairly affluent homes, and 13 of the 26 had their wish granted. 12 of the chosen 13 were boys. Five of the girls who wanted a micro but didn't get one said that their brothers (not necessarily older brothers) had got one for Christmas and that 'sometimes he lets me have a go'. (Gribbin 1984)

This indicates that parental attitudes and expectations have a lot to do with the problem – again, as with mathematics.

Computer comics

A survey carried out by one computer comic (*Load Runner*) (in which it is claimed that boys and girls share equal status) found that the majority of its readers were boys aged 11–13. Many computer comics are particularly stereotyped – designed along the lines of traditional boys' comics – and in newsagents they tend to be displayed next to the BMX magazines etc., at the other end of the shelf to the comics that appeal to girls, so it is not surprising that few girls read them.

The discussion above suggests that parental and consumer pressure has a lot to do with girls' failure to participate equally with boys in computer-related activities, but of course there are other contributory factors. Not only are economic and cultural backgrounds significant, the picture that pupils have of what the computer can offer could also affect their attitudes.

Activity A.3

What is the situation in your school?

(i) Ask the pupils in your tutorial group or one of your classes the following questions:

   (a) What do you like best about using the computer?
   (b) Is there anything that you *don't* like about computers? If so what?
   (c) How relevant do you think that computers will be to you when you leave school?

   Note down the types of comments that are made.

(ii) Jot down up to three ideas of your own about why girls are reluctant to become involved with computers.

(iii) Can you group together your results in any way?

Comment

The anecdotes at the beginning of the appendix provide examples of some of the responses you might receive from your pupils; here are a few more:

> Computers could help you with exams or you could have them at home for playing games on, but computers will be taking over peoples' jobs in the future. That's why people are out of work.
> (Jane, aged 11)

> Well – I suppose that I'm likely to use a word-processor or something when I leave school so computers should be relevant, but there are other things I'd rather be doing at school.
> (Anne, aged 15)

> I think they're the best thing that's happened – much better than TV.
> (Biswajit, aged 14)

The responses from your pupils – and your own views – should have enabled you to write down quite a list of reasons why many girls are reluctant to become involved with computers. We believe that these can probably be grouped together as indicated in figure A.4, although inevitably there is some overlap between these areas.

By examining these three aspects of the problem more closely we hope to suggest ways of improving girls' attitudes towards computers and of increasing their participation in computer-related activities.

Figure A.4 Why girls may be reluctant to become involved with computers

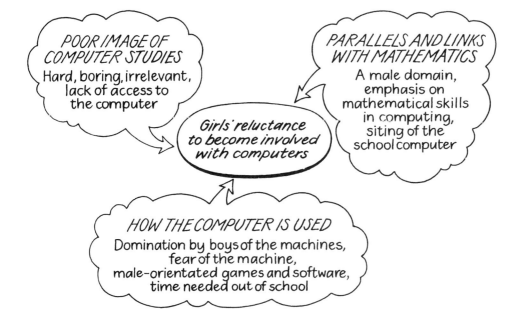

A.2 The poor image of computer studies

Figure A.3 indicates that not only are fewer girls than boys entering public examinations in computer studies and computer science, but that (as with mathematics) the boys tend to perform better at all levels – both in the overall pass rates and more especially in the percentages gaining the higher grades. Actually, the problem is even more acute than this since far too many girls who do opt for computer studies courses at 14 or 15 *drop out* before taking the examination at 16+. For instance, Robin Ward (1986) in *Girls into maths can go* quotes further findings from the Croydon survey mentioned earlier:

> Of 324 Croydon pupils starting a computer studies examination course, completed in the summer of 1982, only 24% were girls, and far too many of those girls (38%), failed to sit the final examination. Some had chosen the subject without knowledge of what the course entailed, were disappointed with the technical aspects and methods of teaching, fell behind and dropped out of the course.

In fact, the survey reported that large proportions of *both* sexes – 47% of the girls and 56% of the boys – were dissatisfied with their courses in computer studies although fewer boys actually opted out. This suggests that the problem of girls' attitudes to, and success in, computer studies lies partly in the nature of the courses that are on offer.

Figure A.5 *Computer studies: an instructional manual* (French 1982)

Contents

Activity A.4

Examine the contents page of a computer studies textbook shown in figure A.5.

(i) Jot down up to three reasons why you think a course covering topics such as those indicated might be inappropriate for fourteen- to fifteen-year-olds.

(ii) Are there any aspects that you feel might cause more problems for girls than for boys?

Comment

We suggest that the contents of this book seem to be very theoretical and are aimed only at children with high ability. Indeed, the author claims that the 'manual aims to satisfy O-level and equivalent computer studies requirements, *and* follows the style of a highly successful A-level computer science text'. Little wonder that most fourteen- to fifteen-year-olds find such courses unappealing!

More generally, figure A.6 shows the reasons for dissatisfaction expressed by the pupils in the Croydon survey, and their concerns regarding computer studies courses may well be reflected in your observations on the contents listed above.

Figure A.6   Reasons given by pupils for dissatisfaction with computer studies courses (EOC 1983; note that multiple responses were taken)

Reason	% Girls	% Boys
Too much writing/theory	17	26
Not enough programming	0	16
Lack of access to computer	20	24
Unspecified boredom	20	6
Poor teaching	10	8
Unfairness in access to machine	3	2
No game-playing	0	5
Difficulty	23	5
Time needed after school	7	0
Mixed-ability class	0	3
Out-of-date material	0	3

Figure A.6 also suggests that there are some differences between the sexes. For example, more girls than boys seem to find computer studies hard or boring and boys appear to want more programming and more games. On the other hand, neither sex likes the amount of writing and theory involved nor the lack of access to machines.

There now appears to be general consent that computer studies syllabuses are inappropriate, so it is not surprising that the subject has a poor image. Unfortunately, many people think of computers and computer studies as being synonymous and the fact that girls tend to have more negative feelings about computer studies than boys must contribute towards their greater aversion to computers themselves in the later years of secondary school. Furthermore, such attitudes are shared by many teachers and this can only make the situation worse!

130

What can be done?

As you may have discovered, younger children who have met the computer only as a learning resource in maths and other classes, or as a tool in a computer appreciation course, do *not* have such negative attitudes, and this suggests one way forward.

(1) Encourage the replacement of computer studies courses at O-level/CSE by computer appreciation courses in the early years of secondary school. Robin Ward (1986) in *Girls into maths can go* writes about one such course developed in Croydon which covers other aspects of information technology (IT) as well as computing:

> The course is specifically designed to interest both sexes, to give all students the opportunity to gain an awareness of and experience with computers, and to gain in confidence and competence without the added burden of an external exam . . . The syllabus was planned to give pupils a wide experience of computers and to show them applications in as many areas as possible. For many, it will be enough to have had the experience of handling and becoming familiar with computers in information technology classes . . .

This information technology approach covers not only computing but also other communication-based topics such as typing, word-processing, printing, coding information as data, keeping records, creating and using data bases and microelectronics. It therefore emphasises the fact that computing is very much concerned with ways of communicating, and this might help to improve girls' attitudes since there is some indication that girls are more interested in – and better at – communication skills than boys.

(2) Too often, pupils have no hands-on experience of computers between leaving primary school and entering the fourth year of secondary school, yet early adolescence is an important period for the development of their ideas. Running a computer appreciation course for twelve- to thirteen-year-olds should capitalise on the enthusiasm expressed by primary-school children and could help girls to retain positive attitudes towards computers.

In addition, if your school wishes to retain its computer studies examination courses, then there are ways of reducing the problems faced by both girls and boys.

(3) Consider the suggestions listed in figure A.7 on page 132. This advice is adapted from that contained in the EOC report *Information technology in schools* (1983). It may be particularly relevant if you have responsibilities for teaching computer studies. Notice that many of these suggestions are similar to those put forward earlier in the pack in relation to improving girls' attitudes to and performance in mathematics.

(4) Exert pressure on Local Authorities and Examining Bodies to change their examination syllabuses for computer studies courses. Such reform is already overdue; many computer studies courses were designed far too quickly as a response to pressure from society and industry for schools to move with the times. Unfortunately the times are still changing, but computer studies courses have yet to catch up!

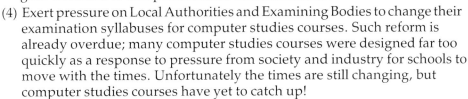

Figure A.7 Ways of improving girls' attitudes towards computer studies courses (adapted from EOC 1983)

What seems to be the problem?	What can be done about it?
(i) Girls opt for computer studies for functional reasons. Few see it as interesting or fun.	Make clear the distinction between the curriculum subject and clubs. Boys too need to realise they will not be playing arcade games. Point out that experience with computer games will not help much at all when it comes to further study or using computers at work.
(ii) Girls are much more favourably inclined to the subject in schools where it is not available.	Examine the image it carries in schools that offer it. Break the link with maths and the elitism.
(iii) Of students who express an interest, more girls fail to take the course. Option patterns deter girls. When girls flounder in choosing options they are less likely to be guided into computer studies.	Confront stereotyped attitudes in staff who: (a) decide option patterns, (b) guide option choice. A major problem seems to be 'blocking' with French.
(iv) Girls tolerate theory and written work more than boys. Girls want more 'hands-on' experience but do not particularly want to program.	Stop apologising for this and make sure boys do not start the course believing it will be all 'practical'. Use packages, data handling, word processors. Be more flexible in classroom management and offer choice in practical work. Consider different approaches to programming.
(v) Girls blame themselves for the difficulty of the subject – boys blame the teacher!	Talk with students about their problems and their progress. Offer quiet consultations with girls rather than trying to 'sort them out' across a class.

A.3   Parallels and links with mathematics

Computers tend to be closely associated with mathematics in two ways. First, in common with other scientific and technological subjects, computing in any form now tends to be regarded as a male-domain, and many of the problems which were discussed earlier in this pack in relation to mathematics apply equally to computers and computing. Second, it is common practice for computers to be housed in maths departments and maths teachers are often actively involved in teaching various aspects of computing. For example, the Croydon survey reported in 1983 that 48% of the computer studies teachers in Croydon held posts in maths departments (Ward 1986 in *Girls into maths can go*). It is true that some maths teachers are coerced into becoming involved with computing against their wishes, but many more *choose* to do so, either because of their personal interests and enthusiasm or because this path offers better chances of promotion. Whether or not this involvement is good for computing – or mathematics – is another matter.

Activity A.5

(i) Compare the reactions of your colleagues to the statement

Mathematics teachers should not have any responsibilities for looking after computer-related activities.

(ii) Jot down a list of points for and against the close ties between mathematics and computers.

Comment

The question of whether or not maths teachers should be involved in computer-related activities usually gives rise to a heated debate with many strong feelings being expressed. There appears to be no simple answer to this question, but the points raised in the following discussion might help you and your colleagues to reach an appropriate conclusion for your school.

Consider first the connection between mathematics and computer studies. On the one hand it is sometimes argued that, because computer studies involves a number of mathematical skills, maths teachers are the most appropriate people to be involved. And indeed there is some sense in the advice 'If you're no good at maths, then don't attempt computer studies', because existing computer studies courses *are* very mathematical – presumably as a result of the involvement of so many mathematicians in drawing up the exam syllabuses. On the other hand, it has already been suggested that such courses are perhaps inappropriate, and there is very little evidence in the real world that prospective employers who use computers expect applicants to have a good grounding in mathematics; they are just as likely to look for qualifications in English or a foreign language – subjects which involve communication skills. In fact many people believe that communication skills are more relevant to computer studies than mathematics and that English teachers (say) might be better qualified to teach the subject than maths teachers who, it is sometimes claimed, are not very good communicators. This line of reasoning suggests that computer studies should be totally divorced from mathematics and it is interesting to note that the Cockcroft report (1982) recommends this course of action. This could have a beneficial effect on both computer studies (by improving its image) *and* mathematics (by leaving maths teachers free to concentrate on mathematics).

However, any conclusion you may reach about the distinction between *computer studies* and mathematics does not necessarily mean that there is no connection at all between mathematics and computing in a more general sense. For example, consider the following extract from *Mathematics from 5 to 16* (DES 1985) about the influence of microcomputers on the approach to the teaching and learning of mathematics.

> . . . the influence of microcomputers on (mathematical) content is likely to increase, and in any case content and approaches react strongly on each other. Microcomputers are a powerful means of doing mathematics extremely quickly and sometimes in a visually dramatic way. If pupils are to use microcomputers in this way they will need to learn to program the machines and so, if programming is not taught elsewhere, it should be included in mathematics lessons. For mathematical purposes such programming does not need to be highly sophisticated. It may be a form of LOGO, or the early stages of a language such as BASIC, or indeed any form of computer control which enables pupils to carry out their own mathematical activities. Having acquired such skill the pupils may use it to study in greater depth

items of mathematical content which are considered important – for example, the properties of numbers, the representation and analysis of data, and the spatial relationships in geometrical figures.

You might like to explore whether any aspects of programming are particularly mathematical. Might the programming language be important and influential?

Using a computer in the way suggested in the extract above emphasises its use as a tool and as a learning resource – as an aid to and a means of learning. As we have already indicated in section A.2, this is likely to encourage positive attitudes towards the machine in *all* pupils and in particular girls. Furthermore, used in the maths classroom as a way of presenting mathematical abstractions in visual form, or as a means of eliminating the drudgery from calculations (as with a calculator), or as a context in which to develop problem-solving skills, the computer could provide a way of improving pupils' attitudes towards mathematics. So perhaps maths teachers *should* indeed take on responsibility for encouraging pupils to develop and use programming skills.

Of course, computers could and should be widely used in other subjects as well, although many people feel that mathematicians tend to regard them as being exclusively for the use of members of the maths department. It is perhaps interesting to note that courses in (say) office practice which tend to have large numbers of girls and which could make considerable use of micros as word-processors tend to be under-resourced with little modern equipment. Often there are just not enough machines to go round, but many schools do have the relevant technology – sitting in the maths department! Indeed, it has been claimed that the 'elitism' and 'aloofness' of some maths teachers *prevents* the widespread use of computers in schools. This suggests that computers should perhaps be regarded simply as a general resource – like a video-recorder – with a resource officer holding the responsibility for their use. If micros are viewed in this way then there could well be more support from all members of staff for the acquisition of more machines if and when resources allow.

What can be done?

A number of possibilities are suggested in the discussion above; these can be summarised as follows:

(1) Encourage the separation of computer studies from mathematics even when the same staff are involved. Emphasise the communication skills aspect of computer studies (which should encourage girls) and, if you have responsibilities for this subject, try to encourage members of staff from other departments to make contributions to the course.

(2) Make more use of the computer as a tool and as a learning resource in maths (and other) classes – including office management – in order to improve the image of both computing *and* mathematics. See the additional information and further reading list at the end of the appendix for references to material that might be particularly appropriate in the maths classroom. Bear in mind though that you will need to take some care in deciding *how* the computer should be used, since it is important that *all* pupils feel that they can use the computer with confidence and assurance (see section A.4).

(3) Put pressure on the authorities to appoint a resources officer for computing
so that the machines can be used by all teachers. Even when
mathematicians do not hide them away it can be very difficult to gain access
to the school computer, and the instruction manuals and the software are
often stored separately – all in all making it very difficult for teachers to use
the computer in class. And teachers often have no-one to turn to for help
when the machines do not work (a frequent occurrence!) since the teacher
in charge of the computer is usually teaching another class. Such
difficulties dissuade teachers from using computers. Appointing a
resources officer or a technician would probably ensure easier access for
everyone and would provide on-the-spot assistance when needed.

A.4   How the computer is used

In this final section we examine the sorts of problems confronted by girls
whenever the computer is used – whether in maths classes, in computer
studies, in computing clubs or even at home.

> Children are arriving early at school and staying behind after lessons to play
> computer games, it was claimed this week. Mr Trevor Fletcher, HMI Inspector with
> special responsibility for mathematics, told the National Union of Teachers'
> National Education Conference on middle schools at Stoke Rochford: 'Children get
> to schools before eight in the morning and don't go home until the caretaker throws
> them out. It's marvellous.'
>
> Of the children getting involved with the microcomputers, boys were more active
> than girls. 'The ribs of the girls are bruised by the boys because they are being
> pushed out of the way in the rush,' Mr Fletcher said . . .   (TES 1983)
>
> . . . And all of the children who did have micros – the few girls as well as the male
> majority – were taught to use it by Dad. None of them ever mentioned Mum using
> it at all – which rather gives the lie to one current TV ad which shows Mum and Dad
> happily playing with the new toy while little Johnny looks on. Typically, I was told
> by my pupils, Dad 'helps the children with the computer in the early evening while
> Mum is cooking the dinner', or at weekends 'while Mum is cleaning up the house'.
> (So) . . . at the age of nine, children in our society are conditioned to accept that boys
> and men are the proper users of a computer, that girls might be allowed an
> occasional touch at the keyboard, and that a woman's job is to feed and care for the
> men.   (Gribbin 1984)

These anecdotes indicate that not only are girls opting out of computing by
choice but that they also experience difficulties in gaining access to the
computer when they *are* interested – either because of the physical domination
by boys of the machines or because of the more subtle pressure exerted by all
walks of society that computers are simply 'not for girls'.

There is plenty of anecdotal evidence that boys dominate computer clubs –
particularly when these are unsupervised. For example, Robin Ward (1983)
reports: 'the boys being larger and louder are apt to take over completely so
that girls feel intimidated to join in; the girls give up eventually – leaving the
field clear for the boys.' Boys are also more likely to dominate the use of the
computer in class unless care is taken to ensure that all pupils have a turn. Do
pupils in your maths class have equal access to the machine? When a group of
pupils is working at the machine can you be sure that the pupil who is actually
working *on* the machine is as likely to be a girl as a boy?

What about the more subtle forms of pressure? The following activity suggests that teachers are as much to blame as any other group in society.

Activity A.6

What are the (approximate) percentages of women and men on the staff of your school who

(i) teach computing;
(ii) use the computer as a learning resource in class;
(iii) have attended an in-service training course on computers;
(iv) use the computer in the staffroom?

Comment

You probably obtained a higher percentage of men than women in all categories, no matter what your school. More men than women teach computing and men often dominate other uses of the computer as well – both in class and in the staffroom. For example, consider the following extract:

> The week before last I ran a workshop at a course for primary teachers who were interested in developing the use of a microcomputer within their school. Approximately one quarter of the teachers there were women, yet I know that a figure more representative of the primary teaching force as a whole would have been 75%.
> . . This week I attended two meetings. One was a regional meeting of advisers and advisory teachers with a responsibility for computing within their particular Local Authority. The other was with a group of HMIs and advisers who were planning a course on the role of the micro in mathematics. At both these meetings I was the only woman in the room.'
> (Straker 1986)

Such discrepancies result partly from the fact that men tend to hold more positions of responsibility than women, but they are also a sign of the greater enthusiasm and confidence shown by men and the fear of machines demonstrated by many women. For example, often when men are asked to solve some technical problem they do not know exactly what to do, but they are prepared to have a go 'by trial and error'. Women very rarely have this level of confidence.

So teachers themselves are *partly* responsible for conveying the impression that computers are designed only to appeal to the male members of our society.

In addition to the difficulties of gaining equal access to the computer in the first place, girls are frequently put off by what it actually does. Many boys view the computer as an extension of an amusement arcade with the associated video-war-type games, and much of the available software is aimed specifically at such interests. You have only to look round the shops at the material designed for the home computer market to verify this for yourself. This problem is also apparent in the educational software available for use in the maths classroom. For example, Brian Hudson (1985) writes:

> On examining a recent catalogue from a company advertising educational software it seemed to me that an assumption had been made that the playing of war games is a wholly healthy pursuit for primary and secondary pupils, particularly in their mathematics lesson.
> For example, a program entitled BATTLES allowed pupils to experience an 'enjoyable game' as they practised plotting coordinates by firing at targets such as submarines and hospitals located on a random basis. To be fair, hospitals qualify for negative scores! A second program rather unadventurously entitled VECTORS

gave the children the refreshing experience of plotting a course, the price of making a mistake being that they would be blown up by a mine. A third program seemed to be based upon the assumption that some experience of blood sports was desirable, perhaps in preparation for the treatment of the human beings that would take place in the war games later! This program was entitled RABBITS and the user's aim was to shoot down up to twenty rabbits appearing in random positions on the screen. Any of these programs might be encountered by quite young children in the primary school. My own four-year-old daughter regards a rabbit as something to marvel over or else to cuddle in bed, and I am sure that even in a few years time she would not derive much pleasure from a game which involved the shooting of creatures she finds so delightful. What also concerns me is the 'macho' male image which the playing of war games undoubtedly has. This I am sure must have a detrimental effect upon the motivation of most girls.

But what sort of computing activities *do* appeal to girls? Charlotte Beyer (1983), reporting on an American survey, suggests that girls prefer stereotyped software titles like 'Typing tutor', 'Typing fractions', 'Consumer buying' and 'Counting calories', while Sherry Turkle (1984) maintains that girls like 'creative' packages which allow them to express their artistic ability.

Activity A.7

What types of computing activities appeal to your pupils?

(i) Give the mini-questionnaire in figure A.8 to your tutorial group or to one of your maths classes (preferably to the same children that you involved in activity A.3).
(ii) Count the number of responses for each sex in each category. Can you identify any patterns? What do your findings in activity A.3 contribute?

(*Note.* Again, you might like to investigate the various responses in different age-groups.)

Comment

Feedback during the development of this pack confirms that boys are likely to play more games than girls, and that girls have a greater preference for word-processor type activities. Furthermore, it was noted that boys often *thought* they were playing a game even if the same activity was labelled as a 'learning activity' by girls. Were your results similar? Did you find that girls tend to prefer the more creative activities? You might like to discuss your findings with your pupils.

What can be done?

There is quite a lot that every maths teacher can do to tackle the problems identified here.

(1) If you are one of those women who has opted out of computing yourself, apply to go on an in-service training course on using the computer in the maths classroom. Encourage other women to do the same.
(2) Ensure that all pupils have equal access to the machine on all occasions.
   Adopt a rota system in the maths classroom. Insist that it is enforced; it might be better, for instance, to leave a machine idle for ten minutes rather than to allow a girl to pass her turn onto someone else.
   If you organise a computer club, make sure that the girls are not put off by the boys. Run separate clubs for girls and boys, or offer different times for different ability levels such as 'expert', 'beginner' and 'complete

Figure A.8 What do you use the computer for?

How many times in the last week have you used the computer:	Never	Once	Sometimes	Frequently
1. To play a game What? _____				
2. To learn something in maths. What? _____ _____				
3. To learn something else. What? _____ _____				
4. As a word-processor. To do what? _____				
5. For some other reason. What? _____ _____				
6. Please tick	Girl ☐	Boy ☐		

beginner', as it has been suggested that girls are more likely to describe themselves as complete beginners than boys. (Distinguishing between pupils in this way also enables provision to be made for the less confident boys as well as providing a means of ensuring equal access for all pupils in single-sex schools.) Ensure that the computer club is supervised; girls are more easily 'pushed out of the way' when there is no teacher present. Feedback during the testing of these materials suggests that girls often experience problems in staying on to the computer club after school since many parents are unwilling for their daughters to go home in the dark (this might apply particularly to girls from some cultural backgrounds). You could therefore ensure that the computer club is as active at lunchtimes as it is after school.

(3) Think carefully about what you use the computer for. Try to ensure that any educational software you use in class appeals to both sexes. Activities which concentrate on the problem solving and investigational aspects of mathematics involving LOGO (say), which is structured and accessible,

seem to be particularly suitable. Our feedback suggests that all pupils who used the computer in this way found the tasks challenging and fun. In fact it was these pupils who exhibited the most positive and enthusiastic attitudes to all aspects of computing, as demonstrated by the following anecdote:

> Computers are very boring, but I like LOGO – it's much better than doing maths on paper! (Salma, aged 11)

Similarly, if you run a computer club, try to include a range of activities which appeal to girls *and* boys – word-processing packages and even LOGO activities, in addition to the usual games. Encourage pupils to use the computer as a tool to (say) write and edit the school magazine. This might provide a good opportunity for girls to demonstrate their better communication skills.

(4) Try to counteract the impression given by parents and the media that computers are not relevant to girls' future careers by challenging the images and assumptions in advertisements etc. Encourage other female members of staff to use the computer whenever appropriate in order to provide good role-models. Chat to your class about their career intentions. Get your pupils to go through the situations vacant column in one of the national papers. How many jobs require the applicant to be familiar with a computer (either with programming skills or with word-processing or information-retrieval skills)? Which types of jobs offer the better salaries? Make available the EOC booklet *Working with computers* which outlines a number of the careers open to women in computing (freely available from the Equal Opportunities Commission). Arrange visits to local industries where computers are widely used. You might be able to obtain a list of suitable firms from your local Trades Council. Banks, manufacturing firms and large supermarkets offer some possibilities.

A.5   Summary and reflection

In this appendix we have investigated (briefly) the various problems faced by girls in connection with computers and computing. Our conclusion is that while it may be highly desirable to divorce computer studies from mathematics, the same cannot be said for all aspects of computing. Indeed, increasing the use of the computer as a learning resource in the maths classroom could *improve* girls' attitudes to computing (with the proviso, of course, that the computer is used sensibly).

We leave you with this final thought: The primary aim of this pack was to suggest action that you can take in the classroom to improve girls' attitudes to and performance in mathematics. Using the computer more *could* have a beneficial effect on *all* pupils' attitudes to mathematics. It can make mathematics more relevant to the real world. It is highly visual and can help children to learn difficult concepts more quickly and at their own pace. If pupils are encouraged to work in pairs then it can be used to promote a more cooperative learning atmosphere in the classroom. It offers tremendous potential for investigative work – particularly with LOGO. Above all, computers can be challenging, exciting and fun. Using them in the right way in mathematics lessons emphasises that maths can also be challenging, exciting and fun – for everybody!

Activity A.8

Jot down in the table below the aspects of the problems discussed in this appendix which are most relevant to your pupils or to your school. Beside each of these, note the strategies you feel you can most effectively introduce in order to improve the situation either now or in the future.

Aspect of the problem	Strategies for tackling it

Further reading and additional information

(1) Articles by R. Ward, A. Straker and M. Gribbin in *Girls into maths can go*

(2) *Information technology in schools* is a very useful publication for anyone wanting to bring about changes in attitudes to school. It is freely available from Equal Opportunities Commission, Overseas House, Quay Street, Manchester M3 3HN.

(3) The EOC also publish the booklet *Working with computers*, which demonstrates the career opportunities for women in computing. It also is available from the above address.

(4) Much of the work in this country on girls and computers has been carried out at the Information Technology Unit, Davidson Centre, Davidson Road, Croydon. In particular you can obtain information about the Croydon course on Information Technology from this address.

(5) For further information on using the microcomputer in the maths classroom you might like to consult the following publications:
- *Micromath*
- SMILE, *Investigator 5*, available from the SMILE Centre, Middle Row School, Kensal Rd, London W10 5DB
- ATM, *Working Notes on Microcomputers in Mathematical Education* and *Some Lessons in Mathematics with a Microcomputer* both available from ATM, King's Chamber, Queen St, Derby DE1 3DA
- Open University, Micros in Schools packs, available from Centre for Continuing Education, The Open University, Milton Keynes MK7 6DH

Bibliography

APU (Assessment and Performance Unit) (1981) *Secondary survey report no 2*, HMSO

ATM (Association of Teachers of Mathematics) (1982) *Geometric images*

ATM (1985) *Working notes on assessment*

Barnes, M., Plaister, R. and Thomas, A. (1984) *Girls count in maths and science*, for the project 'Increasing the participation of girls in mathematics', Commonwealth Schools' Commission, Sydney, Australia

Becker, J. R. (1981) 'Differential treatment of females and males in mathematics classes', *Journal for Research in Mathematics Education*, vol 12, no 1, pp 40–53

*Benett, Y. and Carter, D. (1981) 'Side-tracked – a look at the careers advice given to fifth-form girls' *in* Burton (1986)

Beyer, C. (1983) 'Growing sex gap shows up in computer tastes', *TES*, 18 November 1983

*Binns, B. (1982) 'The girls in my tutor group will not fail at maths' *in* Burton (1986)

*Brown, S. (1984) 'The logic of problem-generation; from morality and solving to de-posing and rebellion' *in* Burton (1986)

Burns, P. F. (1952) *Daily life mathematics*, Book 2, Ginn and Company Ltd

*Burton, L. (ed.) (1986) *Girls into maths can go*, Holt, Rinehart and Winston

*Burton, L. and Townsend, R. (1986) 'Girl friendly schooling' *in* Burton (1986)

Buxton, L. (1981) *Do you panic about mathematics? Coping with maths anxiety*, Heinemann Educational Books

Buxton, L. (1984) *Mathematics for everyman*, Dent

Clift, P. (1978) '. . . And all things nice' (unpublished thesis), Open University School of Education

Clwyd County Council (1983) *Equal opportunities and the secondary school curriculum*, Equal Opportunities Commission

Cockroft, W. H. (chair) (1982) [The Cockroft Report] *Mathematics counts: report of the committee of inquiry into the teaching of mathematics in schools*, HMSO

Cross (chair) (1880) [The Cross report] *Final report of the Commissioners appointed to inquire into the Elementary Education Acts, England and Wales*, HMSO

Dainton, Lord (chair) (1968) [The Dainton Report] *The flow of candidates in science and technology in higher education*, HMSO

Danish Ministry of Education (1983) *Education in Denmark: curriculum regulations for the gymnasium*, Statistiske Efterreting og kultur no 3

DES (Department of Education and Science) (1984) *Statistics of school leavers 1983 volume 2*, HMSO

DES (1985) *Mathematics from 5 to 16 (Curriculum matters 3: an HMI series)*, HMSO

Dweck, C. S. and Bush, E. (1976) 'Sex differences in learned helplessness : I Differential deliberation with peer and adult evaluators', *Developmental Psychology*, vol 12, no 2, pp 147–56

*Eales, A. (1986) 'Girls and mathematics at Oadby Beauchamp College, Leicester' *in* Burton (1986)

Eddowes, M. (1983) *Humble-pi: the mathematics education of girls*, Longmans for the Schools Council

Eddowes, M., Sturgeon, S. and Coates, E. (1980) *Mathematics education and girls*, Sheffield City Polytechnic

Eggleston, J., Galton, M. and Jones, M. (1976) *Processes and products of science teaching (Schools Council Research Studies)*, Macmillan Education

EOC (Equal Opportunities Commission) (1983) *Information technology in schools*, EOC

EOC (1984) *Subject options at school: a positive choice at 13 has a positive effect for life*, EOC

Evans, T. D. (1982) 'Being and becoming: teachers' perceptions of sex-roles and actions towards their male and female pupils', *British Journal of Sociology of Education*, vol 3, no 2, pp 127–43

Everly, B. (1981) *We can do it now, part 1*, EOC

Eynard, R. and Walkerdine, V. (1981) *The practice of reason: investigations into the teaching and learning of mathematics in the early years of schooling vol 2: girls and mathematics*, Thomas Coram Research Unit, University of London

Fennema, E. (1980) 'Sex-related differences in mathematical achievement: where and why' *in* Fox, L., Brody, L. and Tobin, D. (eds.) (1980) *Women and the mathematical mystique*, Johns Hopkins University Press

Fennema, E. and Sherman, J. (1976) 'Sex-related differences in mathematics learning: myths, realities and related factors', paper presented at the American Association for the Advancement of Sciences, Boston USA

French, C. S. (1982) *Computer studies: an instruction manual*, DP Publications

GAIM Project (1985) *Graded Assessment in Mathematics Newsletters*

GAMMA (Gender and Mathematics Association) (1984) *Newsletter 4*, pp 15–17

GCE and CSE boards' joint council for 16+ national criteria (1985) *National criteria for mathematics*, HMSO

Goodwin, S. (1984) 'Women's intuition', *TES*, 22 June 1984

Grady, C. (1983) 'New survey confirms the needs for girls' micro-schemes' *TES*, 16 September 1983

Graf, R. G. and Riddell, J. C. (1972) 'Sex differences in problem-solving as a function of problem context', *Journal of Educational Research*, vol 65, pp 451–2

Graham, A. (1985) *Help your child with maths*, Fontana

Graham, A. and Roberts, H. (1982) *Sums for mums*, Open University

*Gribbin, M. (1984) 'Boys muscle in on the keyboard', *in* Burton (1986)

Griffiths, H. B. and Howson, A. G. (1974) *Mathematics: society and curricula*, Cambridge University Press

Harding, J. (1983) *Switched off: the science education of girls*, Longmans for the Schools Council

Hertfordshire Mathematics Advisory Teachers (1985) *Hertfordshire Mathematics Achievement Certificate*, The Mathematics Centre, Education Centre, Welfield Road, Hatfield

Hudson, B. (1985) 'Please don't shoot the rabbits', *Micromath*, Spring 1985, p 64

ILEA (Inner London Education Authority, Learning Resources Branch) (1984) *Equal Opportunities for Girls and Boys in Primary School Mathematics*

*Isaacson, Z. (1986) 'Freedom and girls' education: a philosophical discussion with particular reference to mathematics' *in* Burton (1986)

Joffe, L. (1983) 'Is it your attitude that matters?', *GAMMA Newsletter*, no 4, pp 7–11

*Joffe, L. and Foxman, D. (1986) 'Attitudes and sex differences: some APU findings' *in* Burton (1986)

Johnstone, B. (1984) 'Technology: must the girls keep losing out?' *The Times*, 6 November 1984

Joshi, H., Layard, R. and Owen, S. (1982) *Female labour supply in post-war Britain*, Centre for Labour Economics, London School of Economics

Kamm, J. (1958) *How different from us: a biography of Miss Buss and Miss Beale*, George Allen and Unwin Ltd

Kant, L. (1982) 'Are examinations up to the mark?' *Secondary Education Journal*, vol 12, no 2, June 1982, pp 8–10

*Kelly, A. *et al.* (1982) 'Gender roles at home and school', *in* Burton (1986)

Kent, D. and Hedger, K. (1980) 'Growing tall', *Educational Studies in Mathematics*, vol 11, pp 137–79

Kindt, M. and de Lange, J. J. Z. N . (1985) *Realistic maths for (almost) all*, for the Hewet Project

Langton, N. and Snape, C. (1984) *A way with maths*, Cambridge University Press

Leder, G. (1980) 'Bright girls, mathematics and fear of success', *Educational Studies in Mathematics*, vol 11, pp 411–21

*Leder, G. (1984) 'Mathematics learning and socialisation processes', *in* Burton (1986)

Ling, J. (1977) *The mathematics curriculum: mathematics across the curriculum*, Blackie

Miller, R. (1981) *Equal opportunities: a careers guide*, Penguin

Millman, V. and Weiner, G. (1985) *Sex differentiation in school – is there really a problem?* Longmans for the Schools Council

Murphy, R. J. L. (1978) *Sex differences in objective test performance*, Report of investigation carried out by the research unit of the Associated Examining Board

*Northam, J. (1982) 'Girls and boys in primary maths books' *in* Burton (1986)

OCEA (1985) *Oxford Certificate of Educational Achievement Newsletters*

Open University, *Micros in Schools*

Osen, L. M. (1974) *Women in mathematics*, MIT

Parr, H. E. (1951) *School mathematics: a unified course*, part 3, G. Bell and Sons Ltd

Payne, G., Hustler, D. and Cuff, T. (1984) *GIST or PIST: teachers' perceptions of the project Girls into Science and Technology*, Manchester Polytechnic

Perl, T. (1978) *Math equals: biographies of women mathematicians and related activities*, Addison Wesley (Mento-Park)

Purves, A. C. and Beach, R. (1972) *Literature and the reader*, final report to the National Endowment for the Humanities, Urbana, Illinois, National Council for Teachers of English

Ricks, F. and Pyke, S. (1973) 'Teachers' perceptions and attitudes that foster or maintain sex-role differences', *Interchange*, vol 4, pp 26–33

Rosenthal, R. and Jacobson, L. (1968) *Pygmalion in the classroom*, Holt, Rinehart and Winston

Russel, S. (1983) *Factors influencing the choice of advanced level mathematics by boys and girls*, Centre for Studies in Science Education, University of Leeds

Schools Council (1983) *Reducing sex differentiation in school*, Newsletter 4, March 1983

*Scott-Hodgetts, R. (1986) 'Girls and mathematics: the negative implications of success' *in* Burton (1986)

Serbin, L. (1978) 'Teachers, peers and play preferences; an environmental approach to sex-typing in the pre-school', *in* Sprung, B. (ed.) (1978) *Perspectives on non-sexist early childhood education*, Teachers College Press

Sharma, S. and Meighan, R. (1980) 'Schooling and sex roles: the case of GCE O-level mathematics', *British Journal of Sociology of Education*, vol 1, no 2, pp 193–205

*Shuard, H. (1986) 'The relative attainment of girls and boys in mathematics in the primary years' *in* Burton (1986)

Shuller, N. (1983) 'Working with images', *Mathematics Teaching*, vol 104, September 1983, pp 38–41

Smail, B. (1984) *Girl-friendly science: avoiding sex bias in the curriculum*, Longmans for the Schools Council

SMILE (1984) *Investigator 1*, SMILE Centre, ILEA

Smith, S. (1983) 'Single-sex setting: a possible solution', text of a lecture given at a conference in Sheffield, July 1982

Smith, M. and Mathew, V. (1984) *Your choice at 13+*, Hobbs Press for the Careers Research and Advisory Centre

SMP (1970) *Book F*, Cambridge University Press

Spender, D. (1980) *Manmade language*, Routledge and Kegan Paul

Spender, D. (1982a) *Invisible women: the schooling scandal*, Writers and Readers

Spender, D. (1982b) *Women's ideas and what men have done to them*, Ark Paperbacks

Stanfield, J. and Potworowska, A. (1971) *Maths adventure*, Evans

*Straker, A. (1986) 'Should Mary have a little computer?' *in* Burton (1986)

Strassberg-Rosenberg, B. and Donlen, T. (1985) 'Content influences on sex differences in performance and aptitude tests', paper presented at the annual meeting of National Council for Measurement in Education, Washington, USA

Swann, M. (chair) (1985) [The Swann Report] *Education for all: report of the Committee of Inquiry into the education of children from ethnic minority groups*, HMSO

Swannell, J. (ed.) (1980) *Oxford dictionary of current English*, Clarendon Press

Sylvester, J. E. K. (1979) *Mainstream mathematics*, book 1, Nelson

*Taylor, H. (1986) 'Experience with a primary school implementing an equal opportunity enquiry', *in* Burton (1986)

Thomas, F. H. (1983) 'An investigation into the relative effects of two methods of teaching geometry', *Mathematical Education for Teaching*, vol 4, no 2, October 1983, pp 19–29

Turkle, S. (1984) *The second self*, Granada

Walden, R. and Walkerdine, V. (1982) *Girls and mathematics: the early years*, University of London Institute of Education, Bedford Way Paper 8

Walden, R, and Walkerdine, V. (1985) *Girls and mathematics: from primary to secondary schooling*, University of London Institute of Education, Bedford Way Paper 24

Ward, M. (1979) *Mathematics and the ten year old*, Evans/Methuen Educational

Ward, R. (1983) 'Wasted potential', *TES*, 4 November 1983

*Ward, R. (1986) 'Girls and technology', *in* Burton (1986)

Weiner, G. (1985) *Just a bunch of girls*, Open University Press

White, W. B. and White, E. (1951) *Essential everyday arithmetic for girls*, part 2, University of London Press

Whyte, J. (1985) *Gender, science and technology: inservice handbook*, SCDC Publications/Longman Resources Unit

Wood, R. (1976) 'Sex differences in mathematics attainment at GCE O-level', *Educational Studies*, vol 2, pp 141–59

Wood, R. (1977) 'Cables' comparison factor: is this where girls' troubles start?', *Mathematics in Schools*, vol 6, no 4, September 1977, pp 18–21

Zimet, S. G. (1976) *Print and prejudice*, Hodder and Stoughton